Food-Borne Viruses

Viruses

Progress and Challenges

Emerging Issues in Food Safety
SERIES EDITOR, Michael P. Doyle

Microbiology of Fresh Produce
Edited by Karl R. Matthews

Microbial Source Tracking
Edited by Jorge W. Santo Domingo and Michael J. Sadowsky

Microbial Risk Analysis in Foods
Edited by Donald W. Schaffner

Enterobacter sakazakii
Edited by Jeffrey M. Farber and Stephen J. Forsythe

Food-Borne Viruses: Progress and Challenges
Edited by Marion P. G. Koopmans, Dean O. Cliver, and
Albert Bosch

ALSO IN THIS SERIES
**Imported Foods: Microbiological Issues
and Challenges (2008)**
Edited by Michael P. Doyle and M. C. Erickson

Food-Borne Viruses

Viruses

Progress and Challenges

EDITED BY

Marion P. G. Koopmans
Laboratory for Infectious Diseases and Prenatal Screening
Centre for Infectious Diseases Control Netherlands
National Institute for Public Health and the Environment
Bilthoven, The Netherlands

Dean O. Cliver
Food Safety Laboratory and World Health Organization
Collaborating Center for Food Virology
Department of Population Health and Reproduction
School of Veterinary Medicine
University of California, Davis
Davis, California

AND

Albert Bosch
Enteric Virus Laboratory
Department of Microbiology
University of Barcelona
Barcelona, Spain

ASM
PRESS

WASHINGTON, DC

Address editorial correspondence to ASM Press, 1752 N St., N.W.,
Washington, DC 20036-2904, USA

Send orders to ASM Press, P.O. Box 605, Herndon, VA 20172, USA
Phone: 800-546-2416; 703-661-1593
Fax: 703-661-1501
E-mail: books@asmusa.org
Online: http://estore.asm.org

Library of Congress Cataloging-in-Publication Data

Food-borne viruses: progress and challenges / edited by Marion P.G.
Koopmans, Dean O. Cliver, and Albert Bosch.
    p. ; cm. — (Emerging issues in food safety)
  Includes bibliographical references and index.
  ISBN-13: 978-1-55581-464-9 (hardcover: alk. paper)
  ISBN-10: 1-55581-464-6 (hardcover: alk. paper) 1. Foodborne diseases.
2. Virus diseases. 3. Food—Microbiology. I. Koopmans, M. P. G. II.
Cliver, Dean O. III. Bosch, Albert. IV. Series.
  [DNLM: 1. Food Microbiology. 2. Food Contamination—prevention &
control. 3. Virus Diseases—prevention & control. QW 85 F6849 2008]

QR201.F62F676 2008
615.9'54—dc22
                                                        2007049165

10 9 8 7 6 5 4 3 2 1

Cover illustration: EM image of a norovirus. (Courtesy of Kjell-Olaf
Hedlund, Swedish Institute for Infectious Disease Control, Solna.)

Contents

Contributors vii
Series Editor's Foreword ix
Preface xi

1 Historic Overview of Food Virology 1
Dean O. Cliver

2 Food-Borne Viruses—State of the Art 29
Marc-Alain Widdowson and Jan Vinjé

3 Enterically Transmitted Hepatitis 65
Rakesh Aggarwal and Sita Naik

4 The Challenge of Estimating the Burden of an Underreported Disease 87
Sarah J. O'Brien

5 Emerging Food-Borne Viral Diseases 117
Erwin Duizer and Marion Koopmans

6 Viral Evolution and Its Relevance for Food-Borne Virus Epidemiology 147
Esteban Domingo and Harry Vennema

7 Rethinking Virus Detection in Food 171
Rosa M. Pintó and Albert Bosch

v

8   Binding and Inactivation of Viruses on and in Food, with a Focus on the Role of the Matrix   189
Françoise S. Le Guyader and Robert L. Atmar

9   Use of the Codex Risk Analysis Framework To Reduce Risks Associated with Viruses in Food   209
Jaap Jansen

10   Risk Assessment of Viruses in Food: Opportunities and Challenges   221
Arie H. Havelaar and Saskia A. Rutjes

Index   237

Contributors

RAKESH AGGARWAL
Department of Gastroenterology, Sanjay Gandhi Postgraduate Institute of
Medical Sciences, Lucknow 226014, India

ROBERT L. ATMAR
Departments of Medicine and Molecular Virology & Microbiology, Baylor
College of Medicine, 1 Baylor Plaza, Houston, TX 77030

ALBERT BOSCH
Enteric Virus Laboratory, Department of Microbiology, University of Barcelona,
08028 Barcelona, Spain

DEAN O. CLIVER
Food Safety Laboratory and World Health Organization, Collaborating Center
for Food Virology, Department of Population Health and Reproduction, School
of Veterinary Medicine, University of California Davis, One Shields Ave.,
Davis, CA 95616-8743

ESTEBAN DOMINGO
Centro de Biología Molecular "Severo Ochoa" (CSIC-UAM), Universidad
Autónoma de Madrid, Cantoblanco, 28049 Madrid, Spain

ERWIN DUIZER
Laboratory for Infectious Diseases, Centre for Infectious Diseases Control,
National Institute for Public Health and the Environment, 3720 BA
Bilthoven, The Netherlands

ARIE H. HAVELAAR
Laboratory for Zoonoses and Environmental Microbiology, Enteric Virus
Laboratory, Department of Microbiology, University of Barcelona, 08028
Barcelona, Spain

JAAP JANSEN
614 Rue Villard, 01220 Divonne les Bains, France

MARION KOOPMANS
Laboratory for Infectious Diseases, Centre for Infectious Diseases Control,
National Institute for Public Health and the Environment, 3720 BA Bilthoven,
The Netherlands

FRANÇOISE S. LE GUYADER
Laboratoire de Microbiologie, Institut Français pour la Recherche et l'Exploitation
de la Mer (IFREMER), Nantes, France

SITA NAIK
Department of Immunology, Sanjay Gandhi Postgraduate Institute of Medical
Sciences, Lucknow 226014, India

SARAH J. O'BRIEN
School of Translational Medicine, University of Manchester, Clinical Sciences
Building, Hope Hospital, Stott Lane, Salford M6 8HD, United Kingdom

ROSA M. PINTÓ
Enteric Virus Laboratory, Department of Microbiology, University of Barcelona,
08028 Barcelona, Spain

SASKIA A. RUTJES
Institute for Risk Assessment Sciences, Division of Veterinary Public Health,
Faculty of Veterinary Medicine, Utrecht University, 3508 TD Utrecht, The
Netherlands

HARRY VENNEMA
Laboratory for Infectious Diseases, Centre for Infectious Diseases Control,
National Institute for Public Health and the Environment, 3720 BA Bilthoven,
The Netherlands

JAN VINJÉ
Respiratory and Enteric Viruses Laboratory Branch, Division of Viral Diseases,
Centers for Disease Control and Prevention, 1600 Clifton Road NE,
Atlanta, GA 30333

MARC-ALAIN WIDDOWSON
Epidemiology Branch, Division of Viral Diseases, Centers for Disease Control and
Prevention, 1600 Clifton Road NE, Atlanta, GA 30333

Series Editor's Foreword

Food-borne viruses, in particular noroviruses, are the most common known causes of food-associated illnesses in much of the world. In the United States alone, public health experts estimate more than 9 million cases of norovirus infections annually. From restaurants to cruise ships, noroviruses have few boundaries, being transmitted mostly by infected humans, especially after they handle food. Although not nearly as prevalent, a variety of other enteric viruses have food-borne disease potential as well and many more are likely to emerge.

Written by a cadre of the world's leading virologists, this book provides state of the science information regarding the evolution and future development of viruses with food-borne potential, the role of viruses in global food-borne illnesses, the challenges and opportunities in developing methods to detect viruses in foods, our current understanding of virus binding on and inactivation in foods, and the application of risk analysis to reducing the risk of food-associated viral illnesses. I know of no other single source that provides such an in-depth, forward-thinking treatise on this subject. It is required reading for anyone interested in not just food-borne viruses but in food-borne disease in general. I commend Marion Koopmans for leading the development of this volume and thank her two collaborators, Dean Cliver and Albert Bosch, for their contributions as Editors. This is truly a benchmark contribution in an area of major significance for the safety of foods.

MICHAEL P. DOYLE, Series Editor
Emerging Issues in Food Safety

Preface

Illness following consumption of food that was contaminated with viruses has been recognized as early as 1914, when four cases of paralytic illness were described among children in an English community who drank raw milk from a common source. That illness was poliomyelitis, later found to be caused by a small virus belonging to a family of viruses that was named *Picornaviridae* (pico = small; RNA is the genome type). Almost a century later, food-borne viral diseases are recognized as a major health concern, but the extent of the problem is poorly defined. Noroviruses and hepatitis A virus, causing vomiting and diarrhea (noroviruses) or liver disease (hepatitis A virus), are the most commonly detected food-borne viruses, and their epidemiology appears to be changing: noroviruses change rapidly over time in a manner similar to influenza A viruses, and increasing levels of hygiene have resulted in an increased susceptibility of populations in high-income countries to hepatitis A that may be imported via food. New viruses are discovered regularly as more is learned about pathogens, and with them new questions arise about the potential for food-borne transmission. The emergence of SARS coronavirus, Nipah virus, and avian influenza viruses from animal reservoirs has illustrated that local food habits may contribute to the spread of pathogens from wild animals to humans and has also shown how difficult it is to determine if food-borne transmission may lead to further dissemination. The global export of foods has more than tripled in the past two decades, resulting in increased risk for large-scale, international outbreaks that are difficult to detect. While regulations are in place to monitor the microbiological quality of food, the criteria in use have been developed based on properties of bacteria, not viruses. Viruses behave quite differently and may remain intact under circumstances in which bacterial contaminants would be killed. In addition, their detection in food requires specialized expertise that is not yet available in most laboratories charged with quality control of food.

What do we know about the recognized food-borne viruses, and what are the gaps? Which lessons can we learn from the past about early detection and control of (emerging) viral infections? What are the challenges in developing reliable ways of detecting if food is contaminated with viruses? What is the role of viral changes through mutation and recombination on their biological properties and epidemiology? These are some of the questions addressed in this book. Chapters have been written by leading scientists in the field, who have been encouraged to provide a challenging in-depth discussion and share their vision for future directions of their field of work. In addition, this book tries to bring scientists and risk managers together by giving a brief background for the methods that have been developed to help decide the best options for controlling food-borne disease and what is needed before these can be used for viral food-borne disease. In short, this book is recommended reading for anyone interested in and/or working on aspects of food-borne viral illness.

Enjoy!

MARION KOOPMANS
DEAN O. CLIVER
ALBERT BOSCH

Food-Borne Viruses: Progress and Challenges
Edited by Marion P. G. Koopmans, Dean O. Cliver, and Albert Bosch
© 2008 ASM Press, Washington, DC

Historic Overview of Food Virology

1

Dean O. Cliver

The development of the microscope led to the discovery of single-celled organisms such as algae, protozoa, and bacteria (14). Although many of these were free-living, environmental forms, it was eventually demonstrated that some of them could cause disease (101). Laboratory methods were developed for cultivation of bacteria, but these generally were not applicable to infectious protozoa. Still, some diseases that were clearly infections rather than intoxications were shown to be caused by agents smaller than bacteria. These infectious agents (*contagium vivum fluidum*) (7) were shown, inter alia, to pass through unglazed porcelain filter candles that would not pass bacteria. An early descriptor was "filterable virus" (165). The word *virus* was originally used in Latin, seemingly to denote a disease agent but perhaps more of a poison than an infectious agent. It evidently was never used in the plural, so modern-day users have been obliged to create their own plural form, most usually *viruses*.

The perception of viruses as having either DNA or RNA but not both, one or more coats of protein, and in some instances a lipid-containing envelope developed over many years. Methods of propagating viruses, first in whole host organisms and then in cultured cells, were essential to progress (57). Electron microscopy provided images of these agents, sometimes biased by technical artifacts. The development of molecular methods for manipulating and characterizing viral nucleic acids led to further great advances and is discussed in detail in chapters 2, 3, and 7. Despite the mass of information that has accumulated, there is a great deal more to be learned. Meanwhile,

DEAN O. CLIVER, Food Safety Laboratory and World Health Organization Collaborating Center for Food Virology, Department of Population Health and Reproduction, School of Veterinary Medicine, University of California, Davis, One Shields Ave., Davis, CA 95616-8743.

the recognition of prions has added an even smaller infectious agent to the picture (145). The vast majority of the approximately 200 human new-variant Creutzfeldt-Jakob disease (vCJD) cases to date have evidently been food-borne, from eating tissues of cattle with bovine spongiform enceph-alopathy (BSE) (58).

I joined the Food Research Institute of the University of Chicago in July of 1962. I had learned cell culture and virology techniques, with a bovine orientation, during my PhD studies at the Ohio State University (29, 30), under the tutelage of an expatriate Hungarian veterinary virologist (95). The founder and Director of the Food Research Institute, Gail M. Dack, invited me to begin a research program on virus transmission via foods. To my knowledge, it was the first such program anywhere, although I soon learned that the U.S. Public Health Service had already been studying virus trans-mission via water (8, 91).

The first recorded outbreak of food-borne disease of probable viral etiol-ogy had occurred long before. In 1914, four cases of paralytic illness occurred in an English community, among children who drank milk from a common source; the mode of contamination of the milk was not determined (90). Nine further outbreaks of food-borne poliomyelitis were reported in the United Kingdom and the United States through 1949, but none were reported thereafter. Diagnoses in these outbreaks were necessarily clinical—1949 was the year that in vitro cultivation of poliovirus was first announced (57)—so laboratory diagnostic methods were not yet available. Raw milk was the apparent vehicle in five of these additional outbreaks (5, 47, 63, 70, 100), and pasteurized milk that perhaps had been contaminated after heating was implicated in two more (118, 128). The only other foods suspected of hav-ing served as vehicles in these outbreaks were lemonade in one instance (143) and cream-filled pastries in another (70). Mosley reviewed eight waterborne outbreaks of poliomyelitis through 1953; six of these occurred in Sweden (135). Raw milk was the recorded vehicle of a hepatitis outbreak comprising 14 cases in Scotland in 1943 (17). What may have been the first recorded shellfish-associated outbreak, comprising more than 600 cases of hepatitis, occurred among people eating raw oysters in Sweden in 1955 to 1956 (146). Most of the earliest known food-borne viral outbreaks were summarized in a 1967 review (23). Various waterborne outbreaks of hepatitis were reviewed by Mosley (135); one that appeared not to have been caused by hepatitis A virus (HAV) (but perhaps was caused by hepatitis E virus) occurred in New Delhi, India, in 1955 to 1956, comprising 28,000 cases (44). Early reviews of food-borne viral diseases were published in 1969 (24) and 1983 (28). The latter document also discusses what was known regarding the agents that were later to be called noroviruses (NoV).

Methods for propagating, detecting, and identifying viruses were in their infancy in the early 1960s (42, 83, 94, 127, 132). Although there had been significant progress in clinical virology, dealing with viruses outside the host required a different mindset. The context and outcomes of the early studies are reviewed in this chapter.

VIRUS PROPAGATION

The earliest approaches to detection of viruses in water and in food were derived from those used in diagnostic clinical virology (83, 94). Although this was reasonable in view of the need to start somewhere, it was soon clear that viral contamination of food and water involved very extensive dilution of the fecal material in which the viruses were shed, so that more sensitive methods were needed (2).

Early detection methods were based on demonstrable infection of a susceptible host system, ideally a cell culture. Monkeys and suckling mice had served in earlier studies of disease diagnosed clinically as paralytic poliomyelitis, but these were very difficult to use in a diagnostic context (131) and less likely still to be of use in food and water virology. Early efforts at in vitro cultivation evidently began with explants of tissue fastened in plasma clots; over time, cells migrated out from the explant into a thin layer in or on the surrounding clotted plasma (132). It was later learned that tissues could be dispersed with trypsin and/or EDTA and the cells could be planted in very clean glass vessels (132). Specialized glassware was made for cell culture, but soft glass prescription bottles were often used (83). The monodisperse cells would attach and form a monolayer, often fortuitously stopping multiplication when growth was confluent on the two-dimensional surface. Further advances included the treatment of plastics (usually polystyrene) to make them compatible with cultured cells (34), the development of CO_2 incubators that could maintain the pH of bicarbonate-buffered culture media in unsealed vessels, the development of completely synthetic media, and in time the establishment of lines of cultured cells that obviated repeated collection of organs to produce primary cultures (42).

Media for primary cell cultures often comprised a balanced salt solution plus an enzymatic hydrolysate of lactalbumin and some blood serum (83). They were compounded from individual ingredients; some media could be sterilized by autoclaving, whereas filters for sterilization were mats of asbestos fibers, which would now be considered hazardous. One of the earliest synthetic media, Medium 199, was apparently compounded to include virtually every biological substance that had been identified in mammalian tissue, including adenine, guanine, thymine, and uracil, whose functions were unknown at the time (134). Eventually, it was shown that established cell lines could be grown and maintained in media in which selected amino acids and vitamins were added to a balanced salt solution plus serum (55). Early handbooks told how to prepare the synthetic media from their individual components (132); fortunately, commercial sources have now obviated this.

Although media now used may be synthetic, they often require blood serum to nurture cultured cells. At one time, homologous (i.e., from the same species as the cells) blood serum was recommended, including collection of human blood for serum to grow human cells (132). More recently, whatever the species of the cells in culture, the serum is most often of bovine (often fetal) origin (158). Those of us who have had the experience of preparing bovine serum for cell culture from blood that we collected, in large containers, from dying cattle in slaughterhouses (29) are grateful that reliable serum is now available from commercial sources.

Against all logic, the organ most frequently used to provide cells and eventually established cell lines for the cultivation of enteric viruses was the kidney. Human kidneys were rarely available, so kidneys of Old World monkeys were the most common source of cells (83). The kidney comprises a rich variety of cells, but cells in culture appear rather uniform. Whether this seeming uniformity results from dedifferentiation or a selective process seems not to have been determined. Only the cortex is used; the dissection of the cortex from the monkey's kidney is straightforward, but the lobed structure of the bovine kidney adds a complication (29). Some of these diverse cells are probably killed during the digestion process, and others do not attach stably to the growing surface and are thus discarded with the first change of medium.

Establishment of a durable line from a primary culture is also an uncertain undertaking: many incipient cell lines are lost through failure to thrive in early passages. A fear of using established cell lines for vaccine production resulted from the example of lines such as HeLa (62), which was derived from a human cervical carcinoma in 1936. It grew readily but, in the absence of precautions, might contaminate and overgrow other cell lines. It was assumed to be malignant and capable of causing cancer if introduced into a human body, and other established lines were suspected of having achieved immunity only by having been converted to malignancy as well. The alternative was production of vaccines from kidneys of monkeys of often-uncertain history, some of which harbored adventitious viruses that were also thought to threaten human health (83). "Live-virus" vaccines, such as the oral polio vaccine devised by Sabin (148), were regarded with more suspicion than vaccines in which the virus had been inactivated with formalin. Still, some of the adventitious viruses of monkeys were apparently more formalin resistant than the polioviruses (151), and some recipients of killed vaccine were exposed to infectious simian virus 40 (144). Although established cell lines are of limited use in detecting the viruses most commonly transmitted to humans via water or food, the U.S. Environmental Protection Agency has standardized a water-testing method that is based on the BGM line of African green monkey (*Cercopithecus aethiops*) cells (60).

Virus replication in cell cultures was often demonstrated with the aid of a microscope: death of host cells maintained in fluid medium was called the cytopathic effect (94). Alternately, viral effects could be localized by maintaining the host cells under semisolid medium, so that killing was confined to zones called plaques (53, 54). Early cell cultures were often produced in prescription bottles (83); this worked reasonably well in plaque assays, but the poor optical properties of the bottles made microscopy difficult. The undulant internal glass surface also affected the distribution of cells in the monolayer and of the inoculum during the adsorption phase of the plaque assay. Because rapid cytopathology would keep the host cells from metabolizing and acidifying their medium, a quantal method was also devised that scored tube cultures or plate wells as negative if the medium was yellow (acidic) and positive if the medium was red (neutral). This scoring by inference was called the metabolic inhibition test (116). Other ingenious applications of cell cultures to the study of enteric viruses, too numerous to list here, had been described before 1970.

Clearly, not all enteric viruses replicate in kidney cells. Cromeans et al. reported that a strain of HAV adapted to replicate and produce plaques in the established FRhK-4 monkey kidney cell line (41), but most strains of HAV lack this capability. Intestinal organ cultures from human fetal material apparently did not support poliovirus replication (49). Replication of coxsackievirus B5 in porcine ileal explants has been reported; this required inclusion of selected hormones (insulin and cortisone) into the CMRL-1066 maintenance medium to keep the explant cultures in a differentiated state (72, 73). Both absorptive and lymphoid tissues were shown to be involved. Murine NoV, an important surrogate for human NoVs in research, replicates in murine macrophage cell line RAW 264.7 (167), which is neither renal nor intestinal. More recently, human NoVs have been shown to replicate, with cytopathology, in a three-dimensional culture of the human embryonic intestinal epithelial cell line INT-407 (158). The cells are grown on microcarriers; although they do not mimic in vivo tissue organization, they do produce visible microvilli. The system appears to offer an excellent research tool—its application to detection of the virus from food or water would be a great advance.

CHARACTERIZATION

Characterization of the viruses of interest was complicated by the lack of laboratory hosts. Landsteiner and colleagues transmitted poliovirus to monkeys as early as 1908, but further studies of monkeys by Flexner and Wollstein led to the erroneous conclusion that the virus infected the monkeys via the intranasal route, presumably by aerosol transmission (131). The transition of poliovirus studies through chimpanzees and eventually to tissue (cell) cultures is ably reviewed by Melnick (131). In the 1940s, newborn mice were tested as a research alternative to monkeys; this led to the discovery of the coxsackieviruses by Dalldorf and Sickles (43). Cultivation of the polioviruses in cell culture by Enders et al. (57) led to testing of fecal specimens in cell culture and the finding of many human enteric viruses that were "orphans" in the sense that they were not immediately associated with illness. The acronym ECHO for "enteric, cytopathogenic, human orphan" was soon adopted (84).

Before infectious hepatitis was designated hepatitis A, there was already an extensive epidemiological record of transmission via food (20). An outbreak of "epidemic jaundice" comprised 69 cases at Yale University (New Haven, CT) in 1921 (82). The investigation was problematic, in that the illness was thought to be caused by a bacterium (*Spirochaeta icterohemorrhagiae*), the incubation period was underestimated, and contact transmission was assumed. Nevertheless, the clinical description closely resembles viral hepatitis (A or E) in humans, onsets were fairly closely clustered (63 within a period of 15 days), and there was strong association with three "eating clubs" that fed a small proportion of the student body. Ten cases of infectious hepatitis occurred in the U.S. state of Georgia in 1945 among people who drank raw milk from an unsanitary, open-air dairy (136). Diagnosis was based on clinical signs; blood from three patients did not reveal antibody to *Leptospira icterohemorrhagiae*. Two earlier cases had been recorded in the area, and the descriptions of feces disposal suggested that there was ample opportunity for contamination of the milk. An outbreak of infectious hepatitis affected 24 general-hospital staff members in 1962 (56). The apparent vehicle was orange juice held at least overnight and perhaps for 36 h before serving, which suggested that the virus was acid stable (the pH of orange juice is generally 3.5 to 4.0). In 1949, Brown (15) reviewed the transmission of virus diseases known to him. On the basis of available epidemiological evidence, he concluded that infectious hepatitis was significantly transmitted by milk and water but there was no evidence for transmission of poliomyelitis by these vehicles.

Among 2,423 outbreaks of food-borne disease reported to the Centers for Disease Control and Prevention during 1988 to 1992, 59% (presumably mostly gastroenteritis) were of undetermined etiology, and an estimated 35% of these were attributable to viruses, along with 4% of outbreaks for which the etiology was determined (6). This would put viral outbreaks at ~26% of the total that occurred and were reported during that period. Of course, not all outbreaks of unknown etiology are caused by viruses, but the proportion attributed to viruses, particularly NoV, has increased greatly in recent years as more states develop the necessary diagnostic capability (see chapter 2). Appleton et al. described outbreaks of gastroenteritis with "long" (24- to 48-h) incubation periods that were surmised probably to be viral on this basis (2). They applied electron microscopy to prepared stool specimens from patients (3) and reported a higher incidence and longer duration of virus detection among infections from shellfish than from other foods. The viruses seen were described as "featureless" and were not detected in extracts of the suspected food vehicles, showing that more sensitive methods were required to detect viruses in food samples.

Eventually, viruses were classified as having RNA or DNA, as having or lacking a lipid-containing envelope, and by size category (83). The nucleic acid type was determined in cell culture: DNA virus replication was inhibited by 5-fluoro-2-deoxyuridine or 5-bromo-2-deoxyuridine in the medium and was often accompanied by production of Feulgen-positive intranuclear inclusions (HCl treatment followed by decolorized rosanaline produces a red color with nucleic acid). The presence of a lipid envelope was surmised if the virus was inactivated by treatment with diethyl ether. Size categories were defined by membrane filtration using filters with pore sizes of 50 and 100 nm. "Small" viruses would pass through the 50-nm filter; "medium" viruses would pass through the 100-nm filter but not 50-nm filter; and "large" viruses were retained by the 100-nm filter. Each of these criteria was strictly qualitative; filtration, for example, resulted in considerable loss of virus titer in passage through the membrane filters (4). It was shown that adsorptive losses of virus during filtration could largely be obviated by incorporating serum into the virus suspension or by pretreating the filter with serum or gelatin (19); later, the tendency of viruses to adsorb to membrane filters of porosity greater than the virion was exploited to concentrate viruses from dilute suspension in water, with subsequent elution in a small volume of liquid (22). In a sense, this was a column chromatography process on a column whose "length" was just the 150-μm thickness of the filter membrane, so the process was first called "membrane chromatography."

Among the many groups of viruses that were identified by nucleic acid type, lipid-containing envelope, and virion size, the enteric viruses were not "large" and did not include lipid-containing envelopes. More sophisticated tests than those summarized above inevitably followed. Nucleic acid type determinations were said to be possible by showing inactivation of a virus by RNase or DNase (166); however, this was surely not applicable to enteric viruses, which must maintain their infectivity in the presence of many extraneous nucleases. Overall, it is difficult to imagine a native virus that would be susceptible to RNase action. In our studies, intact RNA viruses (feline caliciviruses, HAV, and poliovirus) were not affected by RNase, whereas viruses that had been inactivated by heat, chlorine, or UV were attacked by RNase or a combination of proteinase K and RNase (138, 139). Acridine orange staining of infected cells in culture showed orange fluorescence in the presence of single-stranded nucleic acids and yellow-green fluorescence in the presence of double-stranded nucleic acids; it was soon determined that viruses with double-stranded RNA or with single-stranded DNA existed, so the color of fluorescence was not determined on the basis of RNA versus DNA as such (166). Transmission electron microscopy enabled the demonstration of outer membranes and the direct determination of virion size (166), although artifacts were sometimes introduced by the staining or shadowing processes. Rhinoviruses (which cause upper respiratory infections) and enteroviruses were found to be morphologically identical, but the former were inactivated at pH 3, which the latter withstood (69). With increasing knowledge of genome composition and replication strategies, the classification of viruses now exploits these properties as primary criteria for taxonomic purposes. As described in chapters 2, 3, and 6, present-day classification of viruses depends on the nucleotide sequences and organization of the viral genome.

DETECTION AND IDENTIFICATION

An overview of procedures for detecting viruses in food is shown in Fig. 1. Obtaining representative samples for testing represents a great challenge, especially for foods in large lots that cannot be thoroughly mixed before being sampled (see chapter 7). The sample must be liquefied if it is not already a liquid. The earliest methods often involved a mortar and pestle, but mechanical homogenizers were soon more common, with the Stomacher and sonic apparatus as alternatives in some cases (31). Where contamination is likely to be superficial, as on tomatoes, it was reported in 1969 that gentle surface rubbing would dislodge the virus while including a minimum of food solids in the suspension (32). Otherwise, care must be taken that virus is not removed with the food solids when clarifying the suspension.

When our work began in 1962, the obvious clarification methods were centrifugation and filtration. We found that centrifugation worked well with some food suspensions, whereas filtration, even with commercial filter aids,

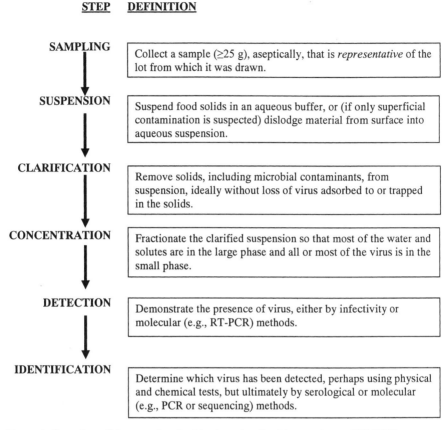

Figure 1 Overview of the steps involved in detecting food-borne viruses. RT-PCR, reverse transcription-PCR.

did not (79). We experimented first with artificially contaminated cottage cheese. We found that homogenization with Freon TF, with an added clay called bentonite, caused food solids to coagulate at the bottom of the tube during centrifugation, leaving the virus in a clear aqueous supernatant. We then adapted the method to suspension and clarification of several other virus-contaminated foods; bentonite was omitted, and serum was added to extract virus from low-protein foods (32, 80). The Freon extraction method not only clarified the samples but also reactivated virus that had been neutralized with coproantibody. Although centrifugation with Freon treatment offered some significant advantages, we were still interested in adapting filtration to the clarification of food sample suspensions. In 1970, we were invited to do some studies on the use of a polycation coagulant, Cat-Floc, in water and wastewater treatment (25); when the project ended, there was a good deal of Cat-Floc left over, so we tested it with virus-inoculated oysters. Recovery of experimentally inoculated virus was 80 to 100%, and the extract obtained could be inoculated directly into cell cultures or concentrated before being tested (110). The method was tested with a wider variety of foods and was found to give at least 80% recovery of virus from meat products, vegetables, baked goods, shellfish, and dairy products (111). From 1979, Cat-Floc was being widely used in recovery of viruses from mollusks (88, 114, 159) and even crabs (156), but the majority of laboratories were clarifying by centrifugation rather than filtration. Although antibiotics are often added to food extracts to be tested in cell cultures, extracts prepared as described above were often sufficiently clear to be passed through membrane filters that would remove bacteria.

Successful clarification yields a minuscule quantity of virus in a relatively large volume of aqueous suspending medium. The task of concentration is to collect as much of the virus as possible in as little of the fluid as possible. In the 1960s, the default concentration method was "brute-force" ultracentrifugation. Any proper virology laboratory had a preparative ultracentrifuge that was capable of imposing forces as great as $100,000 \times g$. Rotors were swinging-bucket, used mostly for density gradient separations (the virus collected on a high-density "cushion" above the bottom of the tube or in between two layers whose specific gravities bracketed the buoyant density of the virus), and preparative (the virus migrates to the point in the tube furthest from the axis of rotation). Preparative rotors were initially made of anodized aluminum and had to be carefully maintained (and the running hours had to be logged) because they were subject to what were euphemistically called "catastrophic disassembly events." When a rotor failed, the potentially deadly pieces were contained within a built-in, cylindrical armor-steel cylinder; the force of the pieces was absorbed by the refrigeration jacket that lined the steel cylinder, while the entire ultracentrifuge moved until it came to rest against a wall. The sound of this event was (I speak from personal experience) disconcerting, even in the next room (my office). Assuming that the virus was pathogenic, the task of disinfecting the debris was daunting. The manufacturer would arrive promptly to collect the pieces for further analysis—photographing the remains was discouraged. Rotors ran in an evacuated chamber to eliminate heating from air friction. When the vacuum failed, as it sometimes did, the rotor would heat up and cook the samples, and heat stress might require de-rating of the rotor.

Even the smallest viruses would form an invisible pellet under these conditions, but as the rotor decelerated at the end of the run, there was some risk that the supernatant would move inertially within the tube and stir some of the virus out of the pellet. At that time, the tubes were made of highly transparent cellulose nitrate (also known as gun cotton), which was both flammable and potentially explosive; any visible pellet in the tube was probably residual food solids rather than the virus. To obviate stir-back loss of virus, we devised a "trap" (38): a 2% gelatin solution is semisolid at 5°C and liquid at room temperature; 0.1 ml of this solution was added to each tube and chilled in place where the pellet would form. The cold virus suspension (food extract) was loaded into the tube, and the tube was placed into the prechilled rotor. At the end of the run, the supernatant was poured off, the gelatin trap was liquefied by application of the hand to the outside of the tube, and the virus was harvested for further processing. The method was quantitatively efficient but still required a preparative ultracentrifuge, with its other hazards and limitations.

Partial concentration of samples, at least to increase the numbers of samples that could be processed in the ultracentrifuge, was accomplished by dialysis against polyethylene glycol (PEG) (21). Water was imbibed from the clarified sample extract, through dialysis tubing, by a concentrated PEG solution or a solid, cast cylinder of PEG (79, 80). Water samples and low-protein food extracts could be concentrated by the membrane chromatography process described above. A "polymer two-phase system" had been devised in Sweden and Denmark, whereby PEG and sodium dextran sulfate plus salt, added directly to the sample suspension, would spontaneously form two phases, with most of the virus being found in the smaller, dextran sulfate phase (122, 142). We found that the dextran sulfate severely interfered with the expression of some viruses in cell culture (65), so we substituted dextran T-500 to obviate this effect (66). More recently, viruses have been precipitated by direct addition of PEG 6000 to the food extract (18, 87). "Organic flocculation" of viruses with beef extract, a method that has been used to elute adsorbed viruses from filter membranes and was first described in 1976 (96), has found many applications in concentrating viruses from environmental and food samples as well (160). Even lettuce floc has been proposed as a means of concentrating virus from dilute suspension (102). Other adsorbents and physical precipitants are also used, and antibody on paramagnetic beads can specifically concentrate selected viruses from dilute suspension (89, 119). Many other methods are used to concentrate clarified food extracts; summarized here are just some of the early methods and their later adaptations.

Detection of viruses in food extracts originally implied the use of primary monkey kidney (PMK) cell cultures, as was also done for diagnostic purposes (83). Large numbers of cultures were often needed, and donor monkeys were typically costly, so that efficient means of producing PMK cultures were much needed. In 1969 we shared suggestions for economical production of PMK cultures (34), and in 1973 we described an apparatus we used for changing the media in large numbers of flask cultures (26). The flask cultures had a monolayer area of 25 cm^2 and were typically inoculated with 0.5 ml of concentrated food extract, often with adsorption for 2 h at room temperature with continuous rocking or for 1 h at 37°C before maintenance medium was added. This gave an inoculum ratio of 20 μl/cm^2. We reported in 1973 that the inoculum ratio could be as great as 1.4 ml/cm^2, with 24-h adsorption at 37°C, if only CPE were sought, with no apparent loss of sensitivity (112). We also showed that, if testing in more than one type of cell culture was required, inoculum could be harvested from one type culture at the end of the adsorption period and tested on another type of culture with little loss of sensitivity. Cytopathic agents isolated from field samples were characterized and identified serologically as described above. Although such procedures are of no use in detecting NoV or HAV, they were significant at that point in the history of food virology.

INACTIVATION

Interest in virus inactivation dates at least as far back as 1884, with respect to rabies virus in rabbit spinal cords (141). Propagation of viruses in cell culture (57) and development of the plaque technique (53, 54) introduced much more quantitative precision into inactivation experiments. It became possible to determine what changes in the virus led to loss of infectivity. In 1967, Dimmock reported extensive studies on the nature of picornavirus inactivation as a function of temperature (46). He found that poliovirus 1 inactivation was thermodynamically different at temperatures above and below 44°C and that, in the higher temperature range, the capsid was degraded. No such degradation was seen at lower temperatures, indicating that it was RNA infectivity that was lost. Breindl reported in 1971 that poliovirus 1 inactivated at 50°C lost all of its VP4 (the smallest of the capsid polypeptides), was unable to attach to host cells, and could be made RNase sensitive by treatment with sodium dodecyl sulfate (13). The viral RNA apparently remained infectious. That same year, Mandel reported that poliovirus 1 had two resonating configurations with isoelectric points (pI) of 7.0 and 4.5 (124). After 5 min at 40°C, the virus was 90% inactivated, and that proportion had a pI of 4.5; low doses of UV radiation did not affect the pI distribution, but high doses shifted the pI to 4.5. The pI 4.5 configuration would appear to relate to the loss of VP4 in Breindl's study, although the temperature is lower than would have been predicted from Dimmock's work. Inactivation is more rapid under mildly alkaline conditions than with acid (152). An elegant 1979 study indicated that poliovirus RNA was inactivated by chlorine and released from the capsid without a conformational change in the capsid (140). There is some evidence that Norwalk virus, and perhaps some of the surrogates for NoV, are relatively resistant to chlorine (52, 97). We showed in 1972 and thereafter that, in addition to chemical and physical means of inactivation, viruses may be inactivated by microbial action in water (33, 81); however, in 1973 we reported that this may not apply to viruses in food, where other substrates for microbial action are present (78). In the 1970s, Konowalchuk and Speirs reported finding various virus inhibitors or inactivators in ground beef (103), fetal bovine serum (103), strawberry extracts (105), grapes and wines (104), tea (107), and apple juice (106, 107). The action of the apple juice was apparently irreversible (106), whereas we found that virus inactivated with grape juice was readily revived with PEG (35). Furthermore, grape juice-inactivated virus and virus neutralized with coproantibody were reactivated by treatment with the contents of porcine stomach and duodenum, suggesting that neither grape juice nor coproantibody would protect against peroral infection (35). Overall, heat, strong oxidizing agents (e.g., chlorine, chlorine dioxide, and ozone), and perhaps UV offer means of inactivating viruses in food; however, there are indications that inactivated virus sometimes tests positive by reverse transcription-PCR (138), so the inactivation processes need to be very well characterized if they are to serve as critical control points for food safety.

LESSONS FROM PRION DISEASES

Now that the focus in virology is firmly fixed on the viral nucleic acid, it is ironic that the latest great food-borne disease threat is from a pathogen that to the best of our knowledge has no nucleic acid (58). A normal prion (PrPC) is a small, glycosylated-protein molecule found mainly in the brain cell membrane; its function remains unknown. The bovine prion consists of 231 amino acid units. An infective prion (PrPSc) has undergone a conformational change and in the process become heat and protease resistant. It can confer this abnormal conformation on another PrPC by contact, assuming that the latter has a susceptible conformation. Unfortunately, bovine PrPSc can affect human PrPC, even though the human prion protein amino acid sequence differs at more than 30 positions from that of the bovine prion protein.

The first food-borne prion disease was kuru, transmitted via cannibalism in New Guinea (61). Scrapie, a prion disease of sheep that had been known for 200 years, is not transmissible to humans, so there was little immediate alarm when BSE ("mad cow disease") was reported in the United Kingdom in 1986—it was assumed to be bovine specific. However, the Spongiform Encephalopathy Advisory Committee was established in 1990 to advise the British government on potential risks (99). Other ruminant species and cats developed spongiform encephalopathies when fed materials from cattle with BSE. The committee advised the government that it was concerned about 10 cases of CJD that had onsets in 1994 and 1995; in 1996, these cases were reported as new-variant CJD (vCJD) in humans. This was very alarming, because unlike classical CJD, which mainly affected the elderly, vCJD principally struck young people.

The BSE epidemic appears to have resulted from a change in the rendering process used in the United Kingdom (164). Because animal fat had become less valuable, the rendering process was changed to conserve energy but yield less fat; the temperature of rendering was reduced and the use of organic solvent to extract the last of the fat was eliminated. A by-product of rendering was meat-and-bone-meal (MBM), which was used in animal feeds. Whether the prions in the MBM originated in cattle or in sheep is unknown, but there were evidently cycles of amplification of BSE prions through cattle, slaughter, rendering, and MBM feeding that raised the BSE incidence in the United Kingdom to an obvious level in the mid-1980s. By 1987, 446 cases of BSE had been recorded in the United Kingdom (http://www.oie.int/eng/info/en_esbru.htm). Research to characterize BSE in cattle was quickly begun. When MBM-containing feeds were banned in the United Kingdom, the material was exported to other countries; as a result, these countries are now experiencing BSE and cases or risk of vCJD.

The BSE infective agent could be detected in the brain, spinal cord, retina, trigeminal and dorsal root ganglia, tonsils, and distal ileum of symptomatic BSE-infected cattle; but extensive tests have failed to detect it in muscle meat or in milk. Potentially positive parts of the animal (specified bovine materials) have been excluded from the food chain in the United Kingdom and, more recently, in the other affected nations. PrP^{Sc} have not been detected in cattle blood, but virus vaccines produced in cell culture for human immunization are increasingly being produced with fetal bovine serum from BSE-free regions, such as Australia and New Zealand, to minimize possible risk.

Of the nearly 200 cases (165 in the United Kingdom) of vCJD reported by the time of this writing, the vast majority are assumed to have been food-borne, although 4 cases in the United Kingdom have now been established as resulting from blood transfusions. Since cooking and other means of inactivating food-borne viruses are not effective against PrP^{Sc}, prevention of transmission via food has depended on excluding infected cattle from the food supply as much as possible and using only portions of the carcass that are unlikely to contain PrP^{Sc} in any event. This strategy seems to have succeeded—the incidence of vCJD has declined in the United Kingdom from 28 in 2000 (the peak year) to 5 in 2006, with further reductions expected. Diagnostic tests have been adapted for possible detection of PrP^{Sc} in food, but these generally lack the needed sensitivity (149). Instead, testing is generally done to detect prohibited central nervous system material (68). Because BSE is seldom seen in young animals, the British government stopped permitting cattle older than 30 months to be eaten; this rule has been relaxed in view of the declining incidence of BSE and of vCJD.

vCJD is a human food-borne disease that is transmitted zoonotically, as are few food-borne viral diseases. Water has not been implicated as a vehicle for PrP transmission. The infectious agent is much smaller than the smallest viruses and contains no nucleic acid. It is resistant to heat that would inactivate viruses in foods and is generally undetectable in edible tissues. The evident success of the control measures now in place suggests that virus transmission via food and water could be prevented in most instances if proper control measures were applied.

PREVENTION

Preventing virus transmission via food or water may be helped by knowing how much virus must be ingested to cause infection (163). Most food-borne viruses are human specific, so this has generally entailed the use of human volunteers. Volunteers were used to estimate the duration of fecal shedding of HAV as early as 1959 (113) and to characterize NoV in feces in 1971 (48). These studies were necessarily qualitative because no laboratory host system was available to quantify the virus, and with human subjects there is a constraint not to cause severe illness. The plaque technique offered a relatively precise and repeatable means of quantifying viruses that would replicate in cell culture (54), although the particle-to-PFU ratio could vary greatly among preparations from the same virus-cell culture combination (74). Weanling pigs, whose digestive tract is very like the human digestive tract, were held in individual isolators and fed human foods until they were challenged perorally in 1975 to 1978 with porcine enteroviruses that were supposed to be virulent (27). One of the experimental subjects and two of the isolation units are shown with the author in Fig. 2. The ratio of cell culture infectivity to peroral infectivity was estimated at roughly 1,000:1, and no illnesses were seen. The ability to maintain the pigs in individual isolators was helpful, but because the animals grew rapidly, their experimental tenure was necessarily limited. In a later trial with echovirus 12 in human volunteers, the infectious dose for 50% of subjects was estimated to be 919 PFU, with 17 PFU estimated to cause infections in 1% of humans (154). Prior infection with echovirus 12 did not reduce susceptibility. A more recent peroral challenge study with NoV has shown that some people are genetically insusceptible (117).

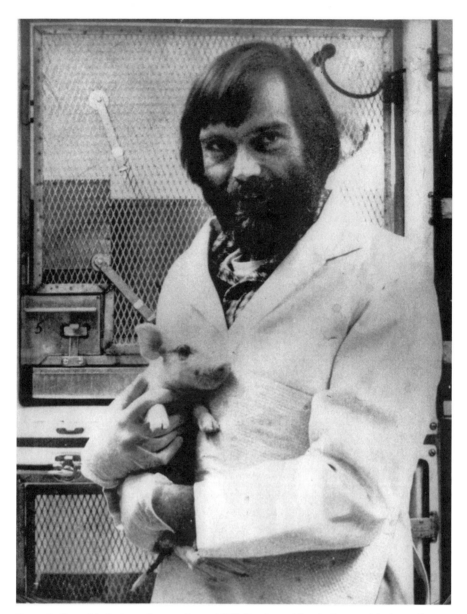

Figure 2 Peroral infectious-dose study, ca. 1977. Porcine subject is shown with the author; two isolation units are visible at the rear.

The most important food-borne viruses emanate from the human intestines and are often flushed down toilets with feces; they are thus introduced into wastewater and may contaminate foods by this route or by direct contact of fecally soiled, unwashed hands with food. If food becomes contaminated, it may be decontaminated by processing or cooking, but the contaminating virus cannot multiply in the vehicle.

Application of vaccines (147, 150) has made poliovirus infections rare in developed countries, and diligent application of HAV vaccines might do the same for hepatitis A (59). At least, hands-on contamination with HAV could be prevented by vaccinating food workers (86).

The antiviral effectiveness of wastewater treatment has been under study for many years. Chlorination of wastewater was being evaluated as early as the 1960s (120), and the need for complete wastewater treatment was recognized in the 1970s (50), in that the viruses in wastewater are often associated with solid particles that mitigate chlorine action (75). The antiviral activity of the microflora in the activated-sludge process was demonstrated in the 1980s (161, 162), after we had shown that microorganisms can inactivate viruses in water (33). It is now clear that virus levels can be greatly reduced by complete wastewater treatment, but in many parts of the world such treatment is not applied before waste or wastewater is discharged (Fig. 3).

Decontamination of hands has been studied to determine its effect on transmission of poliovirus shed naturally in feces (36) and later with HAV grown in cell culture (9). It appeared that virus presence and transfer could be greatly reduced by various cleaning methods. A surprising finding was that rinsing with water, under experimental conditions, was most effective (9).

Assuming that food has become contaminated with virus, many treatments may cause partial or complete inactivation (see chapter 8). Studies of inactivation of viruses in milk began with tick-borne encephalitis virus in the 1950s (64) and continued with enteroviruses in the 1960s (93). The tick-borne encephalitis virus is shed in the milk of infected dairy animals, whereas human enteric viruses are likely to occur in milk only as a result of hand milking with fecally soiled hands. In 2000, we reported that poliovirus 1 was more labile than HAV in milk pasteurization and that the latter showed substantial survival after both low-temperature, long-time and high-temperature, short-time pasteurization (126); comparable results were reported for HAV by Bidawid et al., who also suspended the agent in skim milk and in cream (10). At least one unpublished instance is known to the author, in which there was a real possibility of hands-on contamination of milk with HAV, which was perceived as a significant risk. Meat products were also studied in the 1960s (92), as a result of an outbreak that was recorded in the Soviet Union. In 1973, we reported that coxsackievirus A9, which we had shown to

Figure 3 Fecal-oral transmission cycle in the barrio of Belén, Iquitos, Peru. The photograph shows a family home on a community waterway; the household latrine overhangs the water at left, and a family member is washing dinner dishes in the waterway at right. (photo by Stewart Oakley, California State University, Chico; reprinted with permission.)

be susceptible to proteolytic enzymes (77), was not inactivated during proteolytic spoilage of ground beef or in the fermentative production of Thuringer sausage (78). Lynt reported in 1966 that inoculated enteroviruses persisted well in convenience foods stored at various temperatures and were seemingly unaffected by microbial spoilage (123). He suggested that sodium bisulfite was the ingredient that caused rapid inactivation of coxsackievirus B6 in cole slaw; in 1978, we reported some antiviral activity in sodium bisulfite (153). The 1960s were early times for the U.S. space program; Heidelbaugh and Giron reported considerable inactivation of poliovirus 1 during freeze-drying of foods, as would be done to produce food to feed astronauts, but limited inactivation from cobalt-60 irradiation, a process often applied to space foods (71). In related studies, we showed that enteroviruses persisted well in low-moisture foods, whether the viruses were produced in cell culture or shed in feces (37).

HAV outbreaks associated with frozen strawberries in the United States in the 1990s had attracted special attention to this product (85, 137). Unlike strawberries marketed fresh, strawberries to be frozen can be disinfected before they are subjected to further processing. In 2000, we were asked to evaluate the anti-HAV efficacy of a ClO_2-based disinfection process that was being used commercially on strawberries for freezing. We found that the ClO_2 inactivated only 67% of the HAV, apparently because the solution was recycled repeatedly in the washing-disinfection process, but that the subsequent heating step inactivated 99.98% of the HAV (125). A later study by Gulati et al. yielded unfavorable results with commercial disinfectant products applied to strawberries contaminated with feline calicivirus as a surrogate of NoV (67). Lukasik and colleagues found that various strong oxidants could inactivate poliovirus 1 on strawberries but were best used at much higher levels than recommended by the manufacturers (121).

If viruses are not removed or inactivated by processing, they may still be inactivated by cooking the food. Early attention was devoted to cooking shellfish. As early as 1970, DiGirolamo et al. reported that poliovirus 1 in experimentally contaminated oysters was incompletely inactivated by stewing, frying, baking, and steaming (45). In a 1995 outbreak of hepatitis A from oysters in the United States, those who ate only thoroughly cooked oysters were not at significantly less risk of illness than were raw-oyster eaters (130). HAV was still detectable in mussels 5 min after the shells opened during steaming (1). Therefore, it was not surprising that Croci et al., working with 10^5 PFU of HAV in mussel homogenate, found that complete inactivation occurred after 2 min, but not 1 min, at 100°C (39). In further studies, the group compared three traditional Italian ways of preparing mussels and found that only one of the methods completely inactivated HAV in experimentally contaminated mussels (40). A 1988 outbreak of hepatitis A in the United States involved people who ate sandwiches handled by an infected worker; those who microwaved their sandwiches before eating them were spared (133).

HISTORIC SUMMARY

This brief review has surveyed human viruses and their properties as they relate to transmission via food and water. Early development of the field was limited by the dearth of diagnostic methods for viral diseases, which caused many food-borne outbreaks of illness to be attributed to "undetermined etiology." The development of methods for detecting viruses in water and food and of means (including hazard analysis and critical control point [HACCP] and vaccination) of preventing their transmission via these vehicles has been considered. It is clear that food virology has come of age (108, 109)—at least 10 papers in this field were published during the first half of 2007 (11, 12, 16, 51, 76, 98, 115, 129, 155, 157). Reports of food-borne and waterborne outbreaks are now legion. The remainder of the book provides details of the present and future of food and water virology.

REFERENCES

1. Abad, F. X., R. M. Pinto, R. Gajardo, and A. Bosch. 1997. Viruses in mussels: public health implications and depuration. *J. Food Prot.* **60**:677–681.

2. Appleton, H., S. R. Palmer, and R. J. Gilbert. 1981. Foodborne gastroenteritis of unknown aetiology: a virus infection? *Br. Med. J.* **282**:1801–1802.

3. Appleton, H., and M. S. Pereira. 1977. A possible virus aetiology in outbreaks of food-poisoning from cockles. *Lancet* **i**:780–781.

4. Atoynatan, T., and G. D. Hsiung. 1964. Ultrafiltration of simian viruses. *Proc. Soc. Exp. Biol. Med.* **116**:852–856.

5. Aycock, W. L. 1927. A milk-borne epidemic of poliomyelitis. *Am. J. Hyg.* **7**:791–803.

6. Bean, N. H., J. S. Goulding, C. Lao, and F. J. Angulo. 1996. Surveillance for foodborne-disease outbreaks—United States, 1988–1992. *Morb. Mortal. Wkly. Rep.* **45**:1–66.

7. Beijerinck, M. W. 1964. Concerning a *contagium vivum fluidum* as cause of the spot disease of tobacco leaves, p. 52–63. *In* N. Hahon (ed.), *Selected Papers on Virology*. Prentice Hall Inc., Englewood Cliffs, NJ.

8. Berg, G., N. A. Clarke, and P. W. Kabler. 1962. Interrelationships among ECHO virus types 1, 8, and 12. *J. Bacteriol.* **83**:556–560.

9. Bidawid, S., J. M. Farber, and S. A. Sattar. 2000. Contamination of foods by food handlers: experiments on hepatitis A virus transfer to food and its interruption. *Appl. Environ. Microbiol.* **66**:2759–2763.

10. Bidawid, S., J. M. Farber, S. A. Sattar, and S. Hayward. 2000. Heat inactivation of hepatitis A virus in dairy foods. *J. Food Prot.* **63**:522–528.

11. Botic, T., T. D. Klingberg, H. Weingartl, and A. Cencic. 2007. A novel eukaryotic cell culture model to study antiviral activity of potential probiotic bacteria. *Int. J. Food Microbiol.* **115**:227–234.

12. Boxman, I. L., J. J. Tilburg, N. A. te Loeke, H. Vennema, E. de Boer, and M. Koopmans. 2007. An efficient and rapid method for recovery of norovirus from food associated with outbreaks of gastroenteritis. *J. Food Prot.* **70**:504–508.

13. Breindl, M. 1971. The structure of heated poliovirus particles. *J. Gen. Virol.* **11**:147–156.

14. Brock, T. D. 1961. *Milestones in Microbiology*. Prentice Hall, Englewood Cliffs, NJ.

15. Brown, G. C. 1949. The possible significance of milk and water in the spread of virus infections. *Am. J. Public Health* **39**:764–771.

16. Butot, S., T. Putallaz, and G. Sanchez. 2007. Procedure for rapid concentration and detection of enteric viruses from berries and vegetables. *Appl. Environ. Microbiol.* **73**:186–192.

17. Campbell. 1943. An outbreak of jaundice. *Health Bull.* (Edinburgh) **2**:64.

18. Chung, H., L. A. Jaykus, and M. D. Sobsey. 1996. Detection of human enteric viruses in oysters by in vivo and in vitro amplification of nucleic acids. *Appl. Environ. Microbiol.* **62**:3772–3778.

19. Cliver, D. O. 1965. Factors in the membrane filtration of enteroviruses. *Appl. Microbiol.* **13**:417–425.

20. Cliver, D. O. 1966. Implications of food-borne infectious hepatitis. *Public Health Rep.* **81**:159–165.

21. **Cliver, D. O.** 1967. Detection of enteric viruses by concentration with polyethylene glycol, p. 109–120. *In* G. Berg (ed.), *Transmission of Viruses by the Water Route.* Interscience, New York, NY.

22. **Cliver, D. O.** 1967. Enterovirus detection by membrane chromatography, p. 139–141. *In* G. Berg (ed.), *Transmission of Viruses by the Water Route.* Interscience, New York, NY.

23. **Cliver, D. O.** 1967. Food-associated viruses. *Health Lab. Sci.* **4**:213–221.

24. **Cliver, D. O.** 1969. Viral infections, p. 73–113. *In* H. Riemann (ed.), *Food-Borne Infections and Intoxications.* Academic Press, New York, NY.

25. **Cliver, D. O.** 1971. Viruses in water and wastewater: effects of some treatment methods., p. 149–158. *In* V. Snoeyink and V. Griffin (ed.), *Virus and Water Quality: Occurrence and Control. Proceedings of the Thirteenth Water Quality Conference.* Engineering Publications Office, University of Illinois–Urbana, Urbana.

26. **Cliver, D. O.** 1973. Apparatus for changing tissue culture media, p. 224–226. *In* P. F. Kruse, Jr., and M. K. Patterson, Jr. (ed.), *Tissue Culture: Methods and Applications.* Academic Press, New York, NY.

27. **Cliver, D. O.** 1981. Experimental infection by waterborne enteroviruses. *J. Food Prot.* **44**:861–865.

28. **Cliver, D. O.** 1983. *Manual on Food Virology.* World Health Organization, Geneva, Switzerland.

29. **Cliver, D. O., and E. H. Bohl.** 1962. Isolation of enteroviruses from a herd of dairy cattle. *J. Dairy Sci.* **45**:921–925.

30. **Cliver, D. O., and E. H. Bohl.** 1962. Neutralization of bovine enteroviruses by colostrum, milk and blood serum. *J. Dairy Sci.* **45**:926–932.

31. **Cliver, D. O., R. D. Ellender, and M. D. Sobsey.** 1983. Methods to detect viruses in foods: testing and interpretation of results. *J. Food Prot.* **46**:345–357.

32. **Cliver, D. O., and J. Grindrod.** 1969. Surveillance methods for viruses in foods. *J. Milk Food Technol.* **32**:421–425.

33. **Cliver, D. O., and J. E. Herrmann.** 1972. Proteolytic and microbial inactivation of enteroviruses. *Water Res.* **6**:797–805.

34. **Cliver, D. O., and R. M. Herrmann.** 1969. Economical tissue culture technics. *Health Lab. Sci.* **6**:5–17.

35. **Cliver, D. O., and K. D. Kostenbader, Jr.** 1979. Antiviral effectiveness of grape juice. *J. Food Prot.* **42**:100–104.

36. **Cliver, D. O., and K. D. Kostenbader, Jr.** 1984. Disinfection of virus on hands for prevention of food-borne disease. *Int. J. Food Microbiol.* **1**:75–87.

37. **Cliver, D. O., K. D. Kostenbader, Jr., and M. R. Vallenas.** 1970. Stability of viruses in low moisture foods. *J. Milk Food Technol.* **33**:484–491.

38. **Cliver, D. O., and J. Yeatman.** 1965. Ultracentrifugation in the concentration and detection of enteroviruses. *Appl. Microbiol.* **13**:387–392.

39. **Croci, L., M. Ciccozzi, D. De Medici, S. Di Pasquale, A. Fiore, A. Mele, and L. Toti.** 1999. Inactivation of hepatitis A virus in heat-treated mussels. *J. Appl. Microbiol.* **87**:884–888.

40. **Croci, L., D. De Medici, S. Di Pasquale, and L. Toti.** 2005. Resistance of hepatitis A virus in mussels subjected to different domestic cookings. *Int. J. Food Microbiol.* **105**:139–144.

41. **Cromeans, T., M. D. Sobsey, and H. A. Fields.** 1987. Development of a plaque assay for a cytopathic, rapidly replicating isolate of hepatitis A virus. *J. Med. Virol.* **22**:45–56.

42. **Cunningham, C. H.** 1963. *A Laboratory Guide in Virology,* 5th ed. Burgess Publishing Co., Minneapolis, MN.

43. **Dalldorf, G., and G. M. Sickles.** 1948. An unidentified filtrable agent isolated from the feces of children with paralysis. *Science* **108**:61–62.

44. **Dienstag, J. L.** 1981. Hepatitis A virus: virologic, clinical, and epidemiologic studies. *Hum. Pathol.* **12**:1097–1106.

45. **DiGirolamo, R., J. Liston, and J. R. Matches.** 1970. Survival of virus in chilled, frozen, and processed oysters. *Appl. Microbiol.* **20**:58–63.

46. **Dimmock, N. J.** 1967. Differences between the thermal inactivation of picornaviruses at "high" and "low" temperatures. *Virology* **31**:338–353.

47. **Dingman, J. C.** 1916. Report of possibly milk-borne epidemic of infantile paralysis. *NY State J. Med.* **16**:589–590.

48. **Dolin, R., N. R. Blacklow, H. DuPont, R. F. Buscho, R. G. Wyatt, J. A. Kasel, R. Hornick, and R. M. Chanock.** 1972. Biological properties of Norwalk agent of acute infectious nonbacterial gastroenteritis. *Proc. Soc. Exp. Biol. Med.* **140**:578–583.

49. **Dolin, R., N. R. Blacklow, R. A. Malmgren, and R. M. Chanock.** 1970. Establishment of human fetal intestinal organ cultures for growth of viruses. *J. Infect. Dis.* **122**:227–231.

50. **Dryden, F. D., C. Chen, and M. W. Selna.** 1979. Virus removal in advanced wastewater treatment systems. *J. Water Pollution Control Fed.* **51**:2098–2109.

51. **Dubois, E., C. Hennechart, G. Merle, C. Burger, N. Hmila, S. Ruelle, S. Perelle, and V. Ferre.** 2007. Detection and quantification by real-time RT-PCR of hepatitis A virus from inoculated tap waters, salad vegetables, and soft fruits: characterization of the method performances. *Int. J. Food Microbiol.* **117**:141–149.

52. **Duizer, E., P. Bijkerk, B. Rockx, A. De Groot, F. Twisk, and M. Koopmans.** 2004. Inactivation of caliciviruses. *Appl. Environ. Microbiol.* **70**:4538–4543.

53. **Dulbecco, R.** 1952. Production of plaques in monolayer tissue cultures by single particles of an animal virus. *Proc. Natl. Acad. Sci. USA* **38**:747–752.

54. **Dulbecco, R., and M. Vogt.** 1954. Plaque formation and isolation of pure lines with poliomyelitis viruses. *J. Exp. Med.* **99**:167–182.

55. **Eagle, H.** 1955. Nutrition needs of mammalian cells in tissue culture. *Science* **122**:501–514.

56. **Eisenstein, A. B., R. D. Aach, W. Jacobson, and A. Goldman.** 1963. An epidemic of infectious hepatitis in a general hospital. *JAMA* **185**:171–174.

57. **Enders, J. F., T. H. Weller, and F. C. Robbins.** 1949. Cultivation of the Lansing strain of poliomyelitis virus in cultures of variosu human embryonic tissues. *Science* **109**:85–87.

58. **Erdtman, R., and L. B. Sivitz (ed.).** 2004. *Advancing Prion Science. Guidance for the National Prion Research Program.* National Academies Press, Washington, DC.

59. **Feinstone, S. M.** 1996. Hepatitis A: epidemiology and prevention. *Eur. J. Gastroenterol. Hepatol.* **8**:300–305.

60. **Fout, G. S., F. W. Schaeffer III, J. W. Messer, D. R. Dahling, and R. E. Stetler.** 1996. *ICR Microbial Laboratory Manual.* EPA/600/R-95/178. U.S. Environmental Protection Agency, Office of Research and Development, Washington, DC.

61. Gajdusek, D. C., and V. Zigas. 1957. Degenerative disease of the central nervous system in New Guinea: the endemic occurrence of kuru in the native population. *N. Engl. J. Med.* **257**:974–978.

62. Gey, G. O., and M. K. Gey. 1936. The maintenance of human normal cells and tumor cells in continuous culture. 1. Preliminary report. Cultivation of mesoblastic tumors and normal tissue and notes on methods of cultivation. *Am. J. Cancer* **27**:45–76.

63. Goldstein, D. M., W. M. Hammon, and H. R. Viets. 1946. An outbreak of polioencephalitis among Navy cadets, possibly foodborne. *JAMA* **131**:569–573.

64. Gresikova-Kohutova, M. 1959. Persistence of the virus of tick-borne encephalitis in milk & milk products. *Cesk. Epidemiol. Mikrobiol. Imunol.* **8**:26–32. (In Czech.)

65. Grindrod, J., and D. O. Cliver. 1969. Limitations of the polymer two phase system for detection of viruses. *Arch. Gesante Virusforsch.* **28**:337–347.

66. Grindrod, J., and D. O. Cliver. 1970. A polymer two phase system adapted to virus detection. *Arch. Gesante Virusforsch.* **31**:365–372.

67. Gulati, B. R., P. B. Allwood, C. W. Hedberg, and S. M. Goyal. 2001. Efficacy of commonly used disinfectants for the inactivation of calicivirus on strawberry, lettuce, and a food-contact surface. *J. Food Prot.* **64**:1430–1434.

68. Hajmeer, M., D. O. Cliver, and R. Provost. 2003. Spinal cord tissue detection in comminuted beef: comparison of two immunological methods. *Meat Sci.* **65**:757–763.

69. Hamparian, V. V., M. R. Hilleman, and A. Ketler. 1963. Contributions to characterization and classification of animal viruses. *Proc. Soc. Exp. Biol. Med.* **112**:1040–1050.

70. Hargreaves, E. R. 1949. Epidemiology of poliomyelitis. *Lancet* **i**:972.

71. Heidelbaugh, N. D., and D. J. Giron. 1969. Effect of processing on recovery of poliovirus from inoculated foods. *J. Food Sci.* **34**:239–241.

72. Heinz, B. A., and D. O. Cliver. 1988. Coxsackievirus-cell interactions that initiate infection in porcine ileal explants. *Arch. Virol.* **101**:35–47.

73. Heinz, B. A., D. O. Cliver, and B. Donohoe. 1987. Enterovirus replication in porcine ileal explants. *J. Gen. Virol.* **68**:2495–2499.

74. Heinz, B. A., D. O. Cliver, and G. L. Hehl. 1986. Enumeration of enterovirus particles by scanning electron microscopy. *J. Virol. Methods* **14**:71–83.

75. Hejkal, T. W., F. M. Wellings, A. L. Lewis, and P. A. LaRock. 1981. Distribution of viruses associated with particles in waste water. *Appl. Environ. Microbiol.* **41**:628–634.

76. Hernroth, B., and A. Allard. 2007. The persistence of infectious adenovirus (type 35) in mussels (*Mytilus edulis*) and oysters (*Ostrea edulis*). *Int. J. Food Microbiol.* **113**:296–302.

77. Herrmann, J. E., and D. O. Cliver. 1973. Degradation of coxsackievirus type A9 by proteolytic enzymes. *Infect. Immun.* **7**:513–517.

78. Herrmann, J. E., and D. O. Cliver. 1973. Enterovirus persistence in sausage and ground beef. *J. Milk Food Technol.* **36**:426–428.

79. Herrmann, J. E., and D. O. Cliver. 1968. Food-borne virus: detection in a model system. *Appl. Microbiol.* **16**:595–602.

80. Herrmann, J. E., and D. O. Cliver. 1968. Methods for detecting food-borne enteroviruses. *Appl. Microbiol.* **16**:1564–1569.

81. Herrmann, J. E., K. D. Kostenbader, Jr., and D. O. Cliver. 1974. Persistence of enteroviruses in lake water. *Appl. Microbiol.* **28**:895–896.

82. Hiscock, I. V., and O. F. Rogers, Jr. 1922. Outbreak of epidemic jaundice among college students. *JAMA* **78**:488–490.

83. Hsiung, G. D., and J. R. Henderson. 1964. *Diagnostic Virology*. Yale University Press, New Haven, CT.

84. Hsiung, G. D., and J. L. Melnick. 1955. Plaque formation with poliomyelitis, Coxsackie, and orphan (ECHO) viruses in bottle cultures of monkey epithelial cells. *Virology* **1**:533–535.

85. Hutin, Y. J., V. Pool, E. H. Cramer, O. V. Nainan, J. Weth, I. T. Williams, S. T. Goldstein, K. F. Gensheimer, B. P. Bell, C. N. Shapiro, M. J. Alter, and H. S. Margolis for the National Hepatitis A Investigation Team. 1999. A multistate, foodborne outbreak of hepatitis A. *N. Engl. J. Med.* **340**:595–602.

86. Jacobs, R. J., S. F. Grover, A. S. Meyerhoff, and T. A. Paivana. 2000. Cost effectiveness of vaccinating food service workers against hepatitis A infection. *J. Food Prot.* **63**:768–774.

87. Jaykus, L. A., R. De Leon, and M. D. Sobsey. 1996. A virion concentration method for detection of human enteric viruses in oysters by PCR and oligoprobe hybridization. *Appl. Environ. Microbiol.* **62**:2074–2080.

88. Johnson, K. M., R. C. Cooper, and D. C. Straube. 1981. Procedure for recovery of enteroviruses from the Japanese cockle, *Tapes japonica. Appl. Environ. Microbiol.* **41**:932–935.

89. Jothikumar, N., D. O. Cliver, and T. W. Mariam. 1998. Immunomagnetic capture PCR for rapid concentration and detection of hepatitis A virus from environmental samples. *Appl. Environ. Microbiol.* **64**:504–508.

90. Jubb, G. 1915. A third outbreak of poliovirus at West Kirby. *Lancet* **i**:67.

91. Kabler, P. W., N. A. Clarke, G. Berg, and S. L. Chang. 1961. Viricidal efficiency of disinfectants in water. *Public Health Rep.* **76**:565–570.

92. Kalitina, T. A. 1966. Persistence of coxsackie group B types 3 and 5 in mince meat. *Vopr. Pit.* **5**:74–77. (In Russian.)

93. Kalitina, T. A. 1969. Study of the transmissibility of enteroviruses in milk and milk products. *Zh. Mikrobiol. Epidemiol. Immunobiol.* **7**:61–64. (In Russian.)

94. Kalter, S. S. 1963. *Procedures for the Routine Laboratory Diagnosis of Virus and Rickettsial Diseases.* Burgess Publishing Co., Minneapolis, MN.

95. Kasza, L. 2003. *The Hardships and Joys of an Exiled Cancer Researcher.* 1st Books Library, Bloomington, IN.

96. Katzenelson, E., B. Fattal, and T. Hostovesky. 1976. Organic flocculation: an efficient second-step concentration method for the detection of viruses in tap water. *Appl. Environ. Microbiol.* **32**:638–639.

97. Keswick, B. H., T. K. Satterwhite, P. C. Johnson, H. L. DuPont, S. L. Secor, J. A. Bitsura, G. W. Gary, and J. C. Hoff. 1985. Inactivation of Norwalk virus in drinking water by chlorine. *Appl. Environ. Microbiol.* **50**:261–264.

98. Kingsley, D. H., D. R. Holliman, K. R. Calci, H. Chen, and G. J. Flick. 2007. Inactivation of a norovirus by high-pressure processing. *Appl. Environ. Microbiol.* **73**:581–585.

99. Klingborg, D., and D. Cliver. 2000. Bovine spongiform encephalopathy (BSE), p. 198–204. *In* F. J. Francis (ed.), *The Wiley Encyclopedia of Food Science & Technology,* 2nd ed. John Wiley & Sons, Inc., New York, NY.

100. Knapp, A. C., E. S. Godfrey, Jr., and W. L. Aycock. 1926. An outbreak of poliomyelitis. *JAMA* **87**:635–639.

101. Koch, R. 1961. The etiology of anthrax, based on the life history of *Bacillus anthracis*, p. 89–95. *In* T. D. Brock (ed.), *Milestones in Microbiology*. Prentice Hall, Englewood Cliffs, New Jersey, NJ.

102. Konowalchuk, J., and J. I. Speirs. 1973. Enterovirus recovery with vegetable floc. *Appl. Microbiol.* **26**:505–507.

103. Konowalchuk, J., and J. I. Speirs. 1973. Identification of a viral inhibitor in ground beef. *Can. J. Microbiol.* **19**:177–181.

104. Konowalchuk, J., and J. I. Speirs. 1978. Antiviral effect of commercial juices and beverages. *Appl. Environ. Microbiol.* **35**:1219–1220.

105. Konowalchuk, J., and J. I. Speirs. 1976. Virus inactivation by grapes and wines. *Appl. Environ. Microbiol.* **32**:757–763.

106. Konowalchuk, J., and J. I. Speirs. 1978. Antiviral effect of apple beverages. *Appl. Environ. Microbiol.* **36**:798–801.

107. Konowalchuk, J., and J. I. Speirs. 1976. Antiviral activity of fruit extracts. *J. Food Sci.* **41**:1013–1017.

108. Koopmans, M., and E. Duizer. 2004. Foodborne viruses: an emerging problem. *Int. J. Food Microbiol.* **90**:23–41.

109. Koopmans, M., H. Vennema, H. Heersma, E. van Strien, Y. van Duynhoven, D. Brown, M. Reacher, and B. Lopman. 2003. Early identification of common-source foodborne virus outbreaks in Europe. *Emerg. Infect. Dis.* **9**:1136–1142.

110. Kostenbader, K. D., Jr., and D. O. Cliver. 1972. Polyelectrolyte flocculation as an aid to recovery of enteroviruses from oysters. *Appl. Microbiol.* **24**:540–543.

111. Kostenbader, K. D., Jr., and D. O. Cliver. 1973. Filtration methods for recovering enteroviruses from foods. *Appl. Microbiol.* **26**:149–154.

112. Kostenbader, K. D., Jr., and D. O. Cliver. 1977. Quest for viruses associated with our food supply. *J. Food Sci.* **42**:1253–1257, 1268.

113. Krugman, S., R. Ward, J. P. Giles, O. Bodansky, and A. M. Jacobs. 1959. Infectious hepatitis: detection of virus during the incubation period and in clinically inapparent infection. *N. Engl. J. Med.* **261**:729–734.

114. Landry, E. F., J. M. Vaughn, T. J. Vicale, and R. Mann. 1982. Inefficient accumulation of low levels of monodispersed and feces-associated poliovirus in oysters. *Appl. Environ. Microbiol.* **44**:1362–1369.

115. Leblanc, D., P. Ward, M. J. Gagne, E. Poitras, P. Muller, Y. L. Trottier, C. Simard, and A. Houde. 2007. Presence of hepatitis E virus in a naturally infected swine herd from nursery to slaughter. *Int. J. Food Microbiol.* **117**:160–166.

116. Lennette, E. H., and N. J. Schmidt. 1969. *Diagnostic Procedures for Viral and Rickettsial Infections.*, 4th ed. American Public Health Association, New York, NY.

117. Lindesmith, L., C. Moe, S. Marionneau, N. Ruvoen, X. Jiang, L. Lindblad, P. Stewart, J. LePendu, and R. Baric. 2003. Human susceptibility and resistance to Norwalk virus infection. *Nat. Med.* **9**:548–553.

118. Lipari, M. 1951. A milk-borne poliomyelitis episode. *NY State J. Med.* **51**:362–369.

119. López-Sabater, E. I., M. Y. Deng, and D. O. Cliver. 1997. Magnetic immunoseparation PCR assay (MIPA) for detection of hepatitis A virus (HAV) in American oyster (*Crassostrea virginica*). *Lett. Appl. Microbiol.* **24**:101–104.

120. Lothrop, T. L., and O. J. Sproul. 1969. High-level inactivation of viruses in wastewater by chlorination. *J. Water Pollut. Control Fed.* **41**:567–575.

121. Lukasik, J., M. L. Bradley, T. M. Scott, M. Dea, A. Koo, W. Y. Hsu, J. A. Bartz, and S. R. Farrah. 2003. Reduction of poliovirus 1, bacteriophages, *Salmonella montevideo*, and *Escherichia coli* O157:H7 on strawberries by physical and disinfectant washes. *J. Food Prot.* **66**:188–193.

122. Lund, E., and C. E. Hedstrom. 1966. The use of an aqueous polymer phase system for enterovirus isolations from sewage. *Am. J. Epidemiol.* **84**:287–291.

123. Lynt, R. K., Jr. 1966. Survival and recovery of enterovirus from foods. *Appl. Microbiol.* **14**:218–222.

124. Mandel, B. 1971. Characterization of type poliovirus by electrophoretic analysis. *Virology* **44**:554–568.

125. Mariam, T. W., and D. O. Cliver. 2000. Hepatitis A virus control in strawberry products. *Dairy Food Environ. Sanit.* **20**:612–616.

126. Mariam, T. W., and D. O. Cliver. 2000. Small round coliphages as surrogates for human viruses in process assessment. *Dairy Food Environ. Sanit.* **20**:684–689.

127. Mascoli, C. C., and R. G. Burrell. 1965. *Experimental Virology.* Burgess Publishing Co., Minneapolis, MN.

128. Mathews, F. P. 1949. Poliomeylitis epidemic, possibly milk-borne, in a naval station, Portland, Oregon. *Am. J. Hyg.* **49**:1–7.

129. Mattison, K., K. Karthikeyan, M. Abebe, N. Malik, S. A. Sattar, J. M. Farber, and S. Bidawid. 2007. Survival of calicivirus in foods and on surfaces: experiments with feline calicivirus as a surrogate for norovirus. *J. Food Prot.* **70**:500–503.

130. McDonnell, S., K. B. Kirkland, W. G. Hlady, C. Aristeguieta, R. S. Hopkins, S. S. Monroe, and R. I. Glass. 1997. Failure of cooking to prevent shellfish-associated viral gastroenteritis. *Arch. Intern. Med.* **157**:111–116.

131. Melnick, J. L. 1983. Portraits of viruses: the picornaviruses. *Intervirology* **20**:61–100.

132. Merchant, D. J., R. H. Kahn, and W. H. Murphy, Jr. 1960. *Handbook of Cell and Organ Culture.* Burgess Publishing Co., Minneapolis, MN.

133. Mishu, B., S. C. Hadler, V. A. Boaz, R. H. Hutcheson, J. M. Horan, and W. Schaffner. 1990. Foodborne hepatitis A: evidence that microwaving reduces risk? *J. Infect. Dis.* **162**:655–658.

134. Morgan, J. F., H. J. Morton, and R. C. Parker. 1950. Nutrition of animal cells in tissue culture: initial studies on a synthetic medium. *Proc. Soc. Exp. Biol. Med.* **73**:1–8.

135. Mosley, J. W. 1967. Transmission of diseases by drinking water, p. 5–23. *In* G. Berg (ed.), *Transmission of Viruses by the Water Route.* Interscience, New York, NY.

136. Murphy, W. J., V. M. Petrie, and J. S. D. Work. 1946. Outbreak of infectious hepatitis, apparently milk-borne. *Am. J. Public Health* **36**:169–173.

137. Niu, M. T., L. B. Polish, B. H. Robertson, B. K. Khanna, B. A. Woodruff, C. N. Shapiro, M. A. Miller, J. D. Smith, J. K. Gedrose, M. J. Alter, et al. 1992. Multistate outbreak of hepatitis A associated with frozen strawberries. *J. Infect. Dis.* **166**:518–524.

138. Nuanualsuwan, S., and D. O. Cliver. 2002. Pretreatment to avoid positive RT-PCR results with inactivated viruses. *J. Virol. Methods* **104:**217–225.

139. Nuanualsuwan, S., and D. O. Cliver. 2003. Capsid functions of inactivated human picornaviruses and feline calicivirus. *Appl. Environ. Microbiol.* **69:**350–357.

140. O'Brien, R. T., and J. Newman. 1979. Structural and compositional changes associated with chlorine inactivation of polioviruses. *Appl. Environ. Microbiol.* **38:**1034–1039.

141. Pasteur, L. 1964. A new communication on rabies, p. 30–36. *In* N. Hahon (ed.), *Selected Papers on Virology*. Prentice Hall, Englewood Cliffs, NJ.

142. Philipson, L., P. A. Albertsson, and G. Frick. 1960. The purification and concentration of viruses by aqueous polymerphase systems. *Virology* **11:**553–571.

143. Piszczek, E. A., H. J. Shaughnessy, J. Zichis, and S. O. Levinson. 1941. Acute anterior poliomyelitis: study of an outbreak in West Suburban Cook County, Ill. Preliminary report. *JAMA* **117:**1962–1965.

144. Poulin, D. L., and J. A. DeCaprio. 2006. Is there a role for SV40 in human cancer? *J. Clin. Oncol.* **24:**4356–4365.

145. Prusiner, S. B. 1982. Novel proteinaceous infectious particles cause scrapie. *Science* **216:**136–144.

146. Roos, B. 1956. Hepatitepidemi, spridd genom ostron. *Svenska Lakartidningen* **53:**989–1003.

147. Sabin, A. B. 1965. Oral poliovirus vaccine. History of its development and prospects for eradication of poliomyelitis. *JAMA* **194:**872–876.

148. Sabin, A. B. 1957. Properties of attenuated polioviruses and their behavior in human beings, p. 113–121. *In* V. G. Alfrey (ed.), *Cellular Biology, Nucleic Acids, and Viruses*. New York Academy of Sciences, New York, NY.

149. Safar, J. G., M. Scott, J. Monaghan, C. Deering, S. Didorenko, J. Vergara, H. Ball, G. Legname, E. Leclerc, L. Solforosi, H. Serban, D. Groth, D. R. Burton, S. B. Prusiner, and R. A. Williamson. 2002. Measuring prions causing bovine spongiform encephalopathy or chronic wasting disease by immunoassays and transgenic mice. *Nat. Biotechnol.* **20:**1147–1150.

150. Salk, J. E., P. L. Bazeley, B. L. Bennett, U. Krech, L. J. Lewis, E. N. Ward, and J. S. Youngner. 1954. Studies in human subjects on active immunization against poliomyelitis. II. A practical means for inducing and maintaining antibody formation. *Am. J. Public Health Nations Health* **44:**994–1009.

151. Salk, J. E., U. Krech, J. S. Youngner, B. L. Bennett, L. J. Lewis, and P. L. Bazeley. 1954. Formaldehyde treatment and safety testing of experimental poliomyelitis vaccines. *Am. J. Public Health Nations Health* **44:**563–570.

152. Salo, R. J., and D. O. Cliver. 1976. Effect of acid pH, salts, and temperature on the infectivity and physical integrity of enteroviruses. *Arch. Virol.* **52:**269–282.

153. Salo, R. J., and D. O. Cliver. 1978. Inactivation of enteroviruses by ascorbic acid and sodium bisulfite. *Appl. Environ. Microbiol.* **36:**68–75.

154. Schiff, G. M., G. M. Stefanovic, E. C. Young, D. S. Sander, J. K. Pennekamp, and R. L. Ward. 1984. Studies of echovirus-12 in volunteers: determination of minimal infectious dose and the effect of previous infection on infectious dose. *J. Infect. Dis.* **150:**858–866.

155. Schultz, A. C., P. Saadbye, J. Hoorfar, and B. Norrung. 2007. Comparison of methods for detection of norovirus in oysters. *Int. J. Food Microbiol.* **114:**352–356.

156. Seidel, K. M., S. M. Goyal, V. C. Rao, and J. L. Melnick. 1983. Concentration of rotavirus and enteroviruses from blue crabs (*Callinectes sapidus*). *Appl. Environ. Microbiol.* **46**:1293–1296.

157. Shieh, Y. C., Y. E. Khudyakov, G. Xia, L. M. Ganova-Raeva, F. M. Khambaty, J. W. Woods, J. E. Veazey, M. L. Motes, M. B. Glatzer, S. R. Bialek, and A. E. Fiore. 2007. Molecular confirmation of oysters as the vector for hepatitis A in a 2005 multistate outbreak. *J. Food Prot.* **70**:145–150.

158. Straub, T. M., K. Honer zu Bentrup, P. Orosz-Coghlan, A. Dohnalkova, B. K. Mayer, R. A. Bartholomew, C. O. Valdez, C. J. Bruckner-Lea, C. P. Gerba, M. Abbaszadegan, and C. A. Nickerson. 2007. In vitro cell culture infectivity assay for human noroviruses. *Emerg. Infect. Dis.* **13**:396–403.

159. Vaughn, J. M., E. F. Landry, T. J. Vicale, and M. C. Dahl. 1979. Modified procedure for the recovery of naturally accumulated poliovirus from oysters. *Appl. Environ. Microbiol.* **38**:594–598.

160. Ward, B. K., C. M. Chenoweth, and L. G. Irving. 1982. Recovery of viruses from vegetable surfaces. *Appl. Environ. Microbiol.* **44**:1389–1394.

161. Ward, R. L. 1982. Evidence that microorganisms cause inactivation of viruses in activated sludge. *Appl. Environ. Microbiol.* **43**:1221–1224.

162. Ward, R. L. 1984. Mechanisms of enteric virus inactivation in treatment processes, p. 175–183. *In* J. L. Melnick (ed.), *Monographs in Virology*, vol. 15. *Enteric Viruses in Water. Proceedings of an International Symposium.* S. Karger, Basel, Switzerland.

163. Ward, R. L., and E. W. Akin. 1984. Minimum infective dose of animal viruses. *Crit. Rev. Environ. Control* **14**:297–310.

164. Wilesmith, J. W., J. B. Ryan, and M. J. Atkinson. 1991. Bovine spongiform encephalopathy: epidemiological studies on the origin. *Vet. Rec.* **128**:199–203.

165. Williams, G. 1964. *Virus Hunters.* Alfred A. Knopf, New York, NY.

166. Wilner, B. I. 1969. *A Classification of the Major Groups of Human and Other Animal Viruses*, 4th ed. Burgess Publishing Co., Minneapolis, MN.

167. Wobus, C. E., L. B. Thackray, and H. W. Virgin. 2006. Murine norovirus: a model system to study norovirus biology and pathogenesis. *J. Virol. Methods* **80**:5104–5112.

Food-Borne Viruses: Progress and Challenges
Edited by Marion P. G. Koopmans, Dean O. Cliver, and Albert Bosch
© 2008 ASM Press, Washington, DC

Food-Borne Viruses—State of the Art

2

Marc-Alain Widdowson and Jan Vinjé

Efforts to study and control food-borne disease have traditionally focused on bacterial illness and have overlooked viruses for a variety of reasons. First, sensitive techniques for detecting viruses have been developed only in the last 15 years, whereas bacterial food- and waterborne disease was first described over 100 years ago. Second, viral gastroenteritis has been perceived as a mild, transient illness with no sequelae and for which no specific treatments are available. Third, widespread food and environmental contamination, followed by rapid person-to-person transmission, complicates identification of the contaminated food and implementation of specific control and prevention measures such as food recalls or changes in processing methods. Finally, viral contamination has to date been shown to be of human and not animal origin, so that risk of widespread contamination of animal food products may be lower.

In the last 15 years, however, our understanding of the epidemiology and microbiology of enteric viruses has blossomed, in large part because of the arrival of molecular biology, which has allowed us to diagnose and type viruses. Moreover, as bacterial illnesses become increasingly subject to prevention and control, attention has turned to the remaining causes of food-borne disease. Enteric viral infections are now recognized as a substantial health burden, yet many challenges remain: improving laboratory detection assays, documenting disease burden and transmission, and, most importantly, preventing infection.

MARC-ALAIN WIDDOWSON, Epidemiology Branch, Division of Viral Diseases, Centers for Disease Control and Prevention, Mailstop A34, 1600 Clifton Road NE, Atlanta, GA 30333. JAN VINJÉ, Respiratory and Enteric Viruses Laboratory Branch, Division of Viral Diseases, Centers for Disease Control and Prevention, 1600 Clifton Road NE, Atlanta, GA 30333.

This chapter reviews our current knowledge of food-borne viruses and describes the hurdles to better understanding and ultimate prevention of these infections.

NOROVIRUSES AND EVERYTHING ELSE

Many viruses have the potential to be spread via the food-borne route, including exotic agents such as avian influenza virus and even Ebola virus from consumption of infected apes (see chapter 5). In reality, only a relatively limited number of viruses have been documented as being spread by food. These fall into two main categories: viruses that cause gastrointestinal illness, and those that cause hepatitis. Hepatitis A and E viruses are covered in depth in chapter 3. Many viruses may cause gastroenteritis, but among these noroviruses (NoVs) are by far the most important and frequent cause of food-borne illness. Documented food-borne outbreaks due to other gastroenteritis-causing viruses are rare. Examples include astroviruses, which were implicated in one widespread food-borne outbreak in Japan (127), and rotaviruses, which were implicated in two outbreaks among adults, one associated with consumption of sandwiches at a university in the United States (19) and a second associated with a restaurant in Japan (69). Rotaviruses have been found frequently as contaminants of oysters (98), although no oyster-associated outbreaks of rotavirus infection have been reported. Sapoviruses have rarely been associated with food-borne outbreaks (12, 43). The most likely explanation for the apparent lack of food-borne transmission of most enteric viruses is that they cause infections predominantly among children, whereas adults are rarely affected. Moreover, since children rarely prepare food and are less likely to consume products contaminated in the environment (e.g., rotavirus-contaminated oysters), opportunities for food-borne transmission are far fewer. NoVs, however, are antigenically diverse and may elicit only a short-term immunity; they are therefore common pathogens that cause both child and adult illness and thus have increased opportunity for transmission.

HUMAN CALICIVIRUSES

Through most of the 1940s and 1950s, the role of bacteria as a cause of gastroenteritis was increasingly recognized, yet bacteria could not explain the majority of gastrointestinal illnesses or many large outbreaks of acute-onset vomiting and diarrhea. Volunteer studies demonstrated that infectious agents that could not be cultured remained after stools were filtered for bacteria and that these agents caused acute gastrointestinal illness within 1 or 2 days of ingestion. In 1972, Albert Kapikian employed immune electron microscopy to examine stool samples from ill adult volunteers who had been given bacterium-free filtrates derived from a rectal swab from a child involved in a gastroenteritis outbreak at an elementary school in Norwalk, Ohio (79). This resulted in the first identification of a virus as a cause of gastroenteritis.

The cloning and sequencing of the complete genomes of Norwalk virus and Southampton virus in the early 1990s (70, 95) established these viruses as members of the family *Caliciviridae* in a genus now named *Norovirus* (NoV). In 1977, a virus with a classical calicivirus morphology by electron microscopy was associated with an outbreak of gastroenteritis in an infant home in Sapporo, Japan (26). Cloning and sequencing of this and related viruses from the United Kingdom (104) confirmed that these viruses group into a separate genus now called *Sapovirus* (SaV) (51). Viruses belonging to the other two genera, *Vesivirus* and *Lagovirus*, infect a wide range of animal hosts including cats, dogs, swine, cows, whales, and rabbits (51); a further genus of animal caliciviruses named *Nabovirus* or *Becovirus* has been proposed by some researchers (129, 147).

NoVs are 27- to 32-nm nonenveloped viruses that comprise an icosahedral capsid containing a single-stranded, positive-sense RNA genome. The genome is 7.5 to 7.7 kb in length and contains three open reading frames (ORFs). ORF1 (nucleotides 5 to 5371) encodes a polyprotein of six nonstructural proteins: p48, NTPase, p22, VPg (which binds to the 5' end to initiate translation), $3CL^{Pro}$ (protease), and RdRp (RNA-dependent RNA polymerase). ORF2 (nucleotides 5358 to 6947) encodes the major structural protein (VP1) of approximately 60 kDa that folds into an S (shell) and a P (protruding) domain that is further divided into P1 and P2, with P2 being the most hypervariable region of the genome and being responsible for receptor binding (10, 25). ORF3 (nucleotides 6950 to 7585) encodes a minor structural protein (VP2), the precise function of which is unknown but which has been shown by in vitro studies to upregulate VP1 expression and is involved in the stability of VP1 (10).

NoVs can be classified into five genogroups (GI to GV) based on a >60% amino acid identity in VP1 (51, 164, 180). Human NoVs belong to GI, GII, and GIV, whereas GIII and GV viruses are found only in cattle and mice, respectively. Each genogroup can be further divided into genetic clusters or genotypes based on >80% identity in amino acid sequence of VP1 (51, 164). GI currently contains 8 genotypes, and GII contains 19 genotypes (168, 180), although more genotypes exist based on partial capsid sequences (77). NoVs of genetic cluster GII.4 have predominated in outbreaks throughout the world, whereas GIV viruses are very rare. Although GIII viruses have not been found in human infections, antibodies against these viruses (GIII.2) have been detected in Dutch veterinarians, indicating exposure and possibly infection with bovine NoVs (175). Conversely, several different strains that have been found in pig stools have classified as GII, which contains predominantly human strains (152, 168).

Sapoviruses can be similarly classified into five genogroups, four of which contain viruses detected in humans (41).

DETECTION OF NOROVIRUS IN CLINICAL SAMPLES

Until the mid 1990s, direct electron microscopy was the only method for NoV diagnosis, but for successful detection, direct electron microscopy requires a minimal concentration of 10^6 virus particles per ml of stool (7), which makes this method less suitable for routine diagnosis. Rapid antigen detection assays for the detection of NoV in clinical samples, such as those used for rotavirus detection, are commercially available but as yet do not have the required sensitivity for widespread clinical use. Currently, most laboratories perform TaqMan-based real-time reverse transcription-PCR (RT-PCR) assays for NoV detection. Positive samples are subsequently genotyped by sequencing of RT-PCR products amplified from partial capsid regions.

Conventional RT-PCR Assays

The first NoV RT-PCR assays, developed in the early 1990s (72, 118), were based on a limited number of sequences and were rapidly replaced by second-generation assays that proved to be more broadly reactive and able to detect the majority of the circulating NoV strains. These assays used primer sets targeting the relatively conserved polymerase (region A) region (5, 50, 166), a region at the 3′ end of ORF1 (region B) that is conserved among NoVs of the same genogroup (40), a region (region C) at the 5′ end of ORF2 (50, 77, 88, 125, 164), or a region (region D) at the 3′-end of ORF2, which showed identical phylogenetic grouping to the complete VP1 gene (165). All of these primers have been used successfully in molecular epidemiology studies of NoVs (3, 12, 15, 18, 172), although no single primer set has yet been identified to detect all NoV strains (18, 167). For genotyping, region C sequences have been widely used by clinical diagnostic laboratories in the United States, Europe, and Japan. However, small but important differences, such as a recently detected GII.4 variant strain that cannot be distinguished in this region from previous GII.4 strains, indicate that other capsid regions may need to be considered for accurate typing. The "gold standard" to identify NoV genotypes is full capsid (ORF2) sequence; however, if this is unavailable, region D is recommended for this analysis because sequences in this region have a high correlation with genotyping results derived from complete ORF2 sequences (165).

Real-Time RT-PCR Assays

The ORF1-ORF2 junction region is the most highly conserved region of the NoV genome, containing short stretches of nearly 100% nucleotide identity across strains within a genogroup (76, 82). This region of approximately 100 nucleotides is ideal for designing broadly reactive primers and probes for real-time RT-PCR assays. The first set of assays developed for this region (76) were originally developed as a two-step RT-PCR assay; however, modification into one-step assays (159) has made them more appealing to high-throughput clinical diagnostic laboratories. For clinical samples, the assay detection limit has been reported to be between 10 and 100 copies of RNA (76, 159). Duplexing of the GI and GII assays for the simultaneous detection of both GI and GII NoV has been successful in the two-step format (133) and the modified one-step formats (75, 119). In addition to detecting NoV RNA in clinical samples, the TaqMan real-time RT-PCR assays have been employed to detect NoV RNA in shellfish extracts in France (75, 105) and in water samples (143, 159). While these real-time RT-PCR assays were generally at least as sensitive as nested RT-PCR, discrepancies were observed with some clinical samples (63, 133) and shellfish extracts (105). Such discrepancies are likely to reflect single-nucleotide polymorphisms at primer/probe annealing sites, inhibitors that affect real-time assays more than conventional assays, or degradation after long-term storage of viral RNA. Overall, real-time RT-PCR assays offer rapid and quantitative detection of NoV, not only in clinical specimens but also in food and water samples implicated in NoV transmission during an outbreak.

Enzyme Immunoassays

Rapid assays for antigen detection of NoVs in clinical samples, such as enzyme immunoassays (EIA), offer an attractive alternative to expensive and technically demanding molecular detection assays. However, the development of a broadly reactive EIA for NoVs has been complicated by the large number of antigenically different NoV strains and by the lack of a cell culture system. A major advance came with the production and use of virus-like particles (VLPs) following NoV capsid expression in baculovirus (7, 73), providing antigen for antibody detection and for hyperimmune antiserum production. The first generation of commercial kits included pools of cross-reactive monoclonal and polyclonal antibodies and were evaluated independently by several groups (16, 28, 31, 128, 144). Sensitivities ranged from greater than 70% to lower than 30%, and specificities varied from nearly 50% to 100% in these studies. The general consensus based on these findings is that RT-PCR remains the assay of choice for NoV diagnostics, although the currently commercially available EIA kits may have a role to play in rapid detection of NoVs in outbreak settings, where many stool samples are available for testing. Next-generation assays have been developed and are under evaluation.

Cell Culture and Animal Models

To date, all attempts to propagate human NoV in cell culture have been unsuccessful (36). However, a recent experiment using a three-dimensional cell culture vessel demonstrated for the first time successful passage of both GI and GII NoV in vitro (150). Although this work needs to be confirmed in other laboratories and it will take some time before a routine cell culture assay is available, it is an exciting new avenue for NoV research with important consequences for better detection and prevention methods. Although no small-animal model is available for NoV, human NoVs have been reported to replicate in nonhuman primates and pigs (24, 146, 151). For this reason, most information on the persistence of infectious virus after disinfection, stability in the environment, and basic virology questions of replication and immune correlates of protection has come from human challenge studies and studies using model or surrogate viruses (see chapter 8).

DISEASE PRESENTATION

NoV infection causes acute-onset vomiting and diarrhea, typically after an incubation period of 12 to 48 h. In point-source epidemics, the median incubation period is usually 33 to 36 h. Volunteer studies have demonstrated that as many as 30% of infections may be asymptomatic (48), and one community-based Dutch study found NoV genome in 5% of well controls (30).

Symptoms include vomiting, diarrhea, headaches, nausea, abdominal cramps, and occasionally a low-grade fever. These symptoms can be severe in the acute phase, with repeated vomiting complicating rehydration efforts. As many as 10% of ill persons require medical care, and sometimes brief hospitalization is required to reverse fluid loss (175). In healthy individuals, rapid recovery occurs within 1 to 3 days, although some studies suggest that diarrhea in the elderly or hospitalized takes a few extra days to resolve (110, 114). Long-term sequelae have not been reported, although a recent study from Canada suggests that infection may be associated with a transient postinfectious irritable bowel syndrome that resolves after 3 months (113).

More serious illness has sometimes been attributed to NoV infection. In one example, three military personnel in Afghanistan presented with severe acute gastroenteritis accompanied by obtundation, photophobia, and headache. Two required ventilatory support, and one developed disseminated intravascular coagulation and was evacuated to a military hospital in Germany (20). Possible extraintestinal disease has also been observed in a child who developed encephalopathy after an episode of gastroenteritis (68). NoV RNA was detected in stool, serum, and cerebrospinal fluid by RT-PCR (68). Patients who have received intestinal transplants and subsequent immunosuppressive therapy are at higher risk of more severe disease. These patients can suffer from prolonged chronic NoV diarrhea, which can be confused with allograft rejection (84). Increasing the dose of immunosuppressant to prevent rejection will actually worsen the diarrhea, whereas reduction of therapy has been reported to lead to remission and disappearance of NoV from stools (83).

No deaths directly attributed to NoV have been reported in the literature; however, each year persons in long-term care facilities die with symptoms of gastroenteritis during outbreaks (22). No systematic investigations of deaths in such settings have been performed to determine if NoV was the primary or contributory cause of death. The potential role of NoVs in causing severe disease or death has been poorly investigated due to the lack of a rapid, specific, and sensitive antigen detection assay for use in clinical settings and to the perception that NoV generally causes mild gastroenteritis.

PATHOGENESIS AND IMMUNITY

As few as 10 to 100 ingested viral particles may be sufficient to initiate infection (51). Intestinal biopsies have demonstrated that NoV infects the jejunum, causing blunting of intestinal villi, vacuolation of villous epithelial cells, and mononucleocyte infiltration of the lamina propria but leaving an intact mucosa (2). No lesions have been found in the gastric and colonic mucosa of infected humans (176). The rest of the intestine has not been closely examined in humans, but research using gnotobiotic pigs infected with human GII viruses has found few lesions histologically, and these were confined to the upper intestine (24). Vomiting is likely due to delayed gastric emptying and disrupted motility (117). Interestingly, a transient viremia for 1 to 2 days after inoculation has also been demonstrated in pigs by detection of serum viral antigen and RNA (24). Viremia has not been reported for NoV in humans, although viremia is now recognized as a common occurrence in rotavirus infections among children (13).

The NoV genome can be detected in the stools by RT-PCR for 3 weeks after recovery (145) and possibly longer in infants than in adults (121). Research using quantitative PCR methods such as TaqMan real-time PCR has shown that viral loads may be as much as 100-fold higher in patients with GII NoV infection than in those with G1 infection (23). Moreover, higher viral loads of GII.4 viruses may be associated with more protracted diarrhea and a higher frequency of vomiting (97).

The absence of an infectivity assay means that it has not been possible to determine if prolonged detection of viral RNA signifies viable virus. In turn, this has prevented the development of clear guidelines on how long a person should stay away from contact with others or from handling food after recovery from NoV. However, until such an assay is developed, further work with quantitative PCR methods will help determine the relative viral load in the postrecovery period in severely ill patients compared with those with milder or asymptomatic infection. In addition to providing some data on which to base control guidelines, these studies will also allow a better understanding of the different epidemiology and pathogenesis of strains.

One area of active research is focused on clarifying the apparently complex immunity to NoV infection. One of the first studies with volunteers demonstrated that some subjects developed a short-term immunity of at least 6 to 14 weeks after infection, which correlated with a rise in serum antibody levels. These same individuals were susceptible to illness when rechallenged with the same strain 2 to 3 years later. Other volunteers did not become ill on challenge or rechallenge and did not demonstrate a rise in antibody titers, suggesting innate resistance or acquired immunity (137). Later studies also documented individuals resistant to infection but found that persons with high antibody titers were paradoxically more susceptible when challenged (11, 48, 74), although repeated exposure to the same strain may induce some homologous immunity (74).

In recent years, host genetic factors have been recognized as important determinants of infection and disease, and this work has gone some way to explaining previous conflicting findings. The role of genetic factors was first reported in a study of an outbreak of Norwalk virus in which persons of blood group B were more resistant to infection and to disease than those of blood group O (66). Since this study, substantial effort has been made to better understand this association. Human histo-blood group antigens (HBGA), which are expressed as cellular antigens in gut epithelial cells, are now recognized to be the main sites for NoV binding. This antigen system consists of three main epitopes: A/B, H, and Lewis. Each epitope has different characteristics for binding to different NoV strains, and overall at least seven different binding patterns have been identified (64). For instance, among 77 volunteers challenged with Norwalk virus, the 22 individuals who did not possess the FUT2 gene, which codes for the H type 1 antigen (so-called nonsecretors) did not become infected, and their saliva did not bind to Norwalk virus VLPs. Of the secretors, those of B blood group were at lowest risk of infection, and their saliva either did not bind or bound weakly to

Norwalk virus (103). Different virus genogroups may have different affinities for different receptors. Snow Mountain virus (GII), for instance, does not bind to the H type 1 antigen and has affinity only for the A/B antigen complex (64) (Table 1).

This work raises many new questions about mechanisms of NoV infection, which have widespread implications. First, are there other receptors that play an important role? At least two strains of NoV have been reported not to bind to HBGA, and efforts to culture NoVs have included failed attempts to grow Norwalk virus on cell lines which express H type 1 antigen and therefore should be compatible. Second, what is the NoV carbohydrate-binding site, and how does it interact with host receptors? Although the binding site is known to be in the P2 subdomain, recent work suggests that this domain may form a subviral particle to bind to HBGA (156). Evidence suggests that maternal milk may protect susceptible children from NoV infection by providing decoy receptors, either oligosaccharides or, in the case of secretor mothers, decoy H antigens (71, 101, 157). Understanding these binding mechanisms will help researchers develop certain antiviral drugs that block specific receptors. Third, how do host genetic factors affect the severity of illness? Increased viral load has been associated with both severity of illness and GII.4 NoV strains (23, 97), so certain persons may be susceptible to more severe illness after infection with certain viruses. Fourth, can these interactions explain the epidemiology of NoVs? For instance, GII.4 strains have emerged to be the most prevalent NoVs and have caused increased disease, especially in the winter in closed settings (21, 22, 93, 106). This may be

Table 1 Multiple binding patterns of different NoV strains or VLPs to human HBGAs[a]

Genotype	Strain	Carbohydrate binding[b]								
		H1	H2	H3	Le[a]	Le[b]	Le[x]	Le[y]	A	B
GI.1	Norwalk	+	+	+	−	+	−	+	+	−
GI.3	DSV	−	−	−	−	−	−	−	ND	ND
GII.1	Hawaii	−	ND	−	−	+	ND	ND	+	+ + −[c]
GII.2	SMV	−	ND	−	−	−	ND	ND	−	+
GII.3	Toronto	+	ND	ND	ND	+	ND	ND	+	+
GII.3	Mexico	−	−	−	−	+	−	−	+ +	+
GII.4	Bristol	+	+	ND	−	+	−	−	+	+
GII.4	VA387	+	−	+	−	+	−	+	+	+
GII.5	MOH	+	ND	ND	−	+	−	ND	+	+
GII.9	VA207	−	ND	ND	−	−	+	+	−	−

[a]Data from references 58, 64, and 66.
[b]+, binding; −, no binding; ND, not done.
[c]+ + −, conflicting results.

partly explained by the fact that GII.4 virus strains also have the broadest reactivity, binding to most to the HBGA types (157). HBGA expression varies in different populations around the world and may result in different NoV prevalences. Different susceptibilities of people may result in different viral loads, and transmission may be increased or decreased. Lastly, elucidating the variety of strain-antigen binding patterns will aid our understanding of cross-immunity and will have implications for future vaccine development.

TRANSMISSION

NoVs are transmitted primarily feco-orally, either directly from person to person via contaminated hands or indirectly via contaminated food or water. The virus can also persist for several days to weeks on surfaces and fomites, which can act as a source of continuing infection in outbreaks (see chapter 8). Lastly, airborne droplets of vomitus are infectious and have been strongly implicated in contamination of the environment and in direct transmission of NoV within closed spaces.

Food-borne outbreaks of NoV are often associated with salads, sandwiches, bakery products, and other foods which are handled and contaminated by infected food handlers just before being served. Contaminated oysters have been frequently associated with widespread outbreaks of NoV gastroenteritis (34, 47, 120, 122), but, although these outbreaks have a high profile in the United States, they may represent a minority of food-borne outbreaks compared to contamination at the point of service (175). Contamination of fresh produce at source is also documented as a cause of outbreaks, but interestingly the reports are almost exclusively of contaminated raspberries (27, 39, 44, 61, 91, 100, 142). It remains unclear exactly how the contamination of raspberries occurs, whether in the field or processing, but it is likely that other types of produce that are similarly grown and processed and are eaten raw also pose a risk of NoV infection. Hepatitis A virus, another enteric virus that is resistant to environmental challenge, has been transmitted by several kinds of produce, including green onions and strawberries (65, 169) (see chapter 3).

A recent outbreak investigation in the United States found that frozen, vacuum-packed delicatessen meat had been contaminated with NoV by bare-handed contact during slicing in the meat-processing plant and subsequently was associated with a large outbreak of NoV illness among river rafters (E. J. Barzilay, M. A. Malek, A. Kramer, B. Camp, C. Higgins, L. Jaykus, L. D. Williams, M. Gaither, S. Boone, J. Borajas, M. Lynch, and M.-A. Widdowson, presented at the 5th International Conference on Emerging Infectious Disease, Atlanta, GA, 19 to 22 March 2006). Since bare-handed contact with meat and produce during processing is very common and NoV infections are prevalent, it is likely that disseminated outbreaks are more frequent than currently reported. However, there are several obstacles to a better understanding of how frequently contamination of foods occurs before the point of preparation and service. First, NoV causes a relatively unremarkable illness, so dispersed cases would not be linked clinically. Second, etiologic diagnosis of gastrointestinal illness is often lacking. Third, NoVs detected in stool are frequently not sequenced, limiting the linkage of cases. Lastly, testing of food matrices for NoV is complicated and often not successful (see chapter 7). This is clearly an area where rapid communication of molecular data will uncover potential links between outbreaks or cases due to a common food source. This approach (with pulsed-field gel electrophoretic patterns) has been very successful in uncovering common-source outbreaks of infection due to enteric bacterial agents (154). Networks designed on the same principle have been set up in Europe (Food-Borne Viruses in Europe Network) (89), and a similar initiative is being set up in the United States (CaliciNet). Although these networks have confirmed links between NoV cases with a preexisting epidemiologic connection, they have yet to prove their worth in detecting otherwise unknown links between outbreaks. More extensive, routine, and rapid sequencing and submission of data with subsequent epidemiologic follow-up is required for these systems to achieve their full potential.

Waterborne outbreaks of NoV gastroenteritis in developed countries have been associated with recreational water (62, 141) and with sewage contamination of drinking water. Both groundwater well sources (9) and municipal water systems (14, 94) have been implicated, the latter usually associated with an accompanying chlorination failure.

In addition, NoV-contaminated environments and fomites act as a continuing source of infection (see chapter 8). This characteristic of NoV has been highlighted in the setting of cruise ships, when outbreaks occur on consecutive cruises despite an entirely new cohort of passengers boarding for each voyage and even after 1 week of removal of the ship from service for cleaning (67). Although the crew may act as a source of infection, the rapid increase in the incidence of illness within 1 to 2 days of sailing strongly suggests a common environmental source, especially since most investigations fail to find a common food exposure (Fig. 1). Environmental persistence has also been nicely demonstrated in an outbreak in a theater in which school children, sitting in the same seats as children from a different school the day before who had vomited on those seats, went on to develop illness due to the same NoV strain (38). Public toilets, even where soiling is unapparent, have been implicated in the transmission of NoV on an airplane (172).

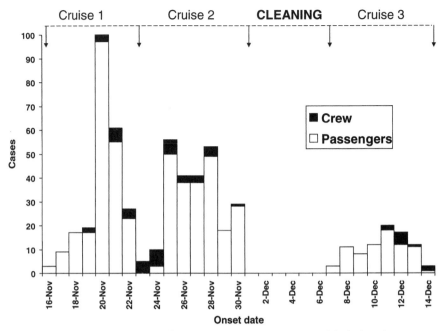

Figure 1 Number of passengers and crew reporting acute gastroenteritis during three consecutive cruises on one ship, 16 November to 14 December 2002 (*n* = 575). (Reprinted from reference 67.)

A further mode of transmission, peculiar to NoV, is via droplets of vomitus. Similar to a flushed toilet, which generates a powerful aerosol and may contaminate the environment (46), vomiting produces droplets which settle on surfaces and possibly are ingested by people in the proximity. One report of an outbreak in a banquet hall convincingly showed an increasing risk of illness the closer someone was seated to the index patient who vomited unexpectedly while dining (111).

Many outbreaks are propagated by several of these modes of transmission, and a lack of understanding of which mode predominates at any one time poses a severe challenge to the implementation of focused control measures. Outbreaks in closed settings, such as nursing homes, hospitals, and cruise ships, are especially difficult to control because the virus spreads easily from person to person. Environmental testing is occasionally used to document contamination, but the methods are relatively insensitive, and since it is not feasible to swab all possibly contaminated surfaces, a negative finding is no assurance of lack of contamination.

Molecular Epidemiology

Depending on the year, about 75 to 95% of NoV outbreaks worldwide are due to GII viruses and 5 to 25% are due to GI viruses, with GIV viruses being identified only very rarely. However, strain diversity varies from year to year in different countries. Since 1995, when the U.S. 95/96 GII.4 variant was reported as globally emergent (125), most studies of NoV molecular diversity have reported GII.4 strains as predominant on each continent (6, 15, 18, 106, 123, 140). Within the GII.4 clade, however, specific strains are regularly replaced by other GII.4 strains, which emerge to cause substantial increases in outbreak frequency (see chapter 6). In 2002 and 2004, a GII.4 strain caused a sharp rise in the frequency of outbreaks in closed settings such as hospitals and nursing homes in Europe (106) and nursing homes and cruise ships in the United States (21, 171). Again, in 2006 and 2007, closely related GII.4 variants were implicated in a large increase in NoV activity in nursing homes in both Europe and the United States (22, 93). In the United States, the sequence on the capsid region suggests that during each period of emergence, variants of one particular substrain will predominate (Fig. 2). It remains unclear why strains belonging to the GII.4 cluster predominate and, moreover, why they seem to be associated with outbreaks in closed settings rather than being transmitted by food (12). Factors previously discussed include higher viral loads in GII.4 infection (23), the possibility that older persons in nursing homes may shed more virus (97), and broader receptor-binding patterns of GII.4 viruses. Other possibilities include different physicochemical properties leading to increased virus persistence in the environment or a different symptomology during infection, such as increased vomiting and therefore more disseminated and rapid spread. One puzzle is why these GII.4 viral lineages have predominated for almost 10 years now and have not led to some level of herd immunity that would predispose to the emergence of other strains not recently encountered by immune systems. It may be that the small changes in nucleotide sequences between these very closely related viruses translate as critical differences in immunogenic antigens (see chapter 6).

DISEASE BURDEN

Sporadic Disease

The cloning of the Norwalk virus genome allowed the development of sensitive RT-PCR detection methods and serologic and antigen detection assays, which used VLPs or monoclonal antibodies derived from animals immunized to specific VLPs. Cross-sectional serologic studies have shown

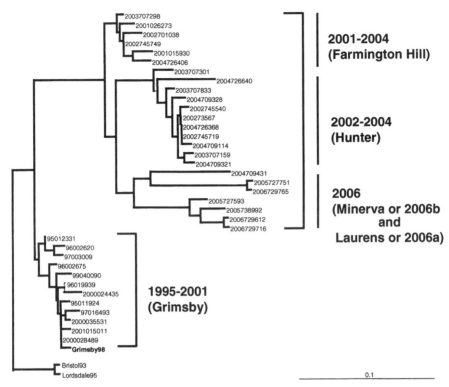

Figure 2 Phylogenetic analysis of NoV GII.4 strains detected in stool samples from outbreaks in the United States, 1995 to 2006, demonstrating four different subclades or variants. A multiple alignment was created using ClustalW, and a neighbor-joining tree was generated on the basis of uncorrected distances of a 277-bp region at the 5′ end of ORF2. Lordsdale95, Bristol93, Grimsby95, and Farmington Hill (FMH03) are included as GII.4 reference strains.

that NoVs are highly ubiquitous, with evidence of infection in all populations, including isolated Amazonian Indian groups (42). Serologic studies using a wide variety of VLPs have found that prevalences of antibodies to GI and GII NoVs rise quickly in the first few years of life. For instance, a study of over 2,000 serum specimens from individuals from four southern African countries found a seroprevalence of over 90% of antibodies to both Norwalk virus (GI) and Mexico virus (GII) among children younger than 5 years (148). High prevalences of NoV antibodies also have been reported from developed countries (60, 139, 174), including among young children (102, 139). Several studies show a trend for anti-Norwalk virus (GI) antibodies to be developed later than antibodies against GII viruses (49, 102, 132, 139). This may reflect the finding that in the last 10 to 15 years, for which reliable strain typing data are available, GI virus infections have been less common than GII virus infections. Some work has attempted to identify possible risk factors for antibody acquisition and has suggested that low socioeconomic status (53, 132) and poorer hygiene (138) may be associated with seropositivity.

One major limitation of serologic studies, however, is that it is not possible to translate antibody titers into levels of protective immunity. Although seroprevalences to NoVs reach over 90% by age 5, unlike other gastrointestinal viruses such as rotaviruses, adults remain susceptible to infection and clinical disease with NoVs.

More recently, RT-PCR assays have been used to better characterize the epidemiology of endemic NoV illness in children and adults both in the community and in health care settings (see chapter 9). A population-based study in The Netherlands provided an age-standardized NoV prevalence estimate of 11% among persons with gastroenteritis of any severity at all ages, with a slightly decreased prevalence in adults (30). Other population-based studies have found similar results: 7% in the United Kingdom and 11% in Australia, although neither estimate was age standardized (112, 170). One recent study in the United States, however, found NoVs in fewer than 2% of children younger than 3 years with mild diarrhea (163).

The prevalence of NoV in stools of children younger than 5 years hospitalized for diarrhea varies widely. Most studies find NoV in 5 to 20% of children admitted for diarrhea (33, 56, 57, 87, 126, 131, 134, 140), although one study in Peru found that up to 35% of hospitalized children were positive for NoV in stool by PCR and a liquid hybridization assay (136). Fewer efforts have focused on the role of NoVs as a cause of severe disease among adults, where, in general, fewer episodes of diarrhea requiring health care are associated with NoV compared with infants. Among persons older than 14 years with diarrhea who entered a hospital in Hong Kong, 8.2% were positive for NoV by RT-PCR (96), but in South Africa NoVs were associated with only 2.3% of cases of diarrhea among patients older than 19 years (178).

However, the results of studies looking at the role of NoV in gastroenteritis can be difficult to interpret for several reasons. First, antigen detection assays and even some RT-PCR assays may not detect the range of possible NoV types. Second, specimen collections are often unrepresentative of diarrheal illness from all causes, since case definitions vary and specimens that have tested positive for bacteria are often screened out before being tested for viruses. Lastly and most importantly, the frequent lack of data on mixed infections and on controls challenges straightforward attribution of illness to NoV infection. NoVs have been found in 1 to 12% of well controls (30, 131, 136) by sensitive RT-PCR methods. The goal, therefore, remains to design and implement longitudinal controlled studies to investigate the contribution of NoV to severe diarrhea of any cause in persons of all ages. Part of that challenge is to apply laboratory tests that will help discriminate between persons with clinical NoV infection and persons with incidental NoV present in the stool. This may require a reassessment of the role of the traditional RT-PCR and a shift to quantitative PCR, which may prove useful in trying to determine a threshold NoV viral load of clinical significance. Antigen detection assays are overall less sensitive than molecular assays in the detection of low viral loads, but they may nonetheless be sensitive enough to detect higher viral loads of NoV associated with disease. In particular, there is a lack of studies to estimate the disease burden in certain populations, such as the elderly, the infirm, hospitalized patients, or residents of long-term care facilities, where outbreak reports suggest that disease frequency is elevated and illness may result in death (22). Similarly, further well-designed studies in developing countries will help us understand the global disease burden of NoV, which may be increased in poorer populations. For instance, deaths due to rotavirus infection are very rare in rich countries with high levels of health care (173); however, globally, rotavirus is implicated in more than 500,000 deaths, the large majority in the poorest countries of south Asia and sub-Saharan Africa (135).

Lastly, determining the proportion of sporadic NoV disease that may be food borne is challenging and relies on interpretation of outbreak data. Outbreak surveillance systems and the data collected vary considerably (see the next section), and therefore recent studies aimed at estimating the incidence of food-borne disease and the contribution of different pathogens have reached disparate conclusions. In the United States, NoV is estimated to cause 12% of all 76 million cases of food-borne gastroenteritis a year (116). In Australia this percentage is 8% (54), and in the United Kingdom it is only 3% (1). Nevertheless, in all studies NoVs were the most or second most frequent known cause of food-borne disease.

Outbreaks

In addition to sporadic disease, NoV infection often causes large outbreaks of acute gastroenteritis among persons of all ages, characterized by high attack rates, secondary transmission, illness with a short incubation period (12 to 48 h) and duration (median 24 to 72 h), and a high prevalence of vomiting (>50% of cases). In the absence of laboratory tests, these clinical and epidemiologic criteria have been found to be highly specific in determining whether an outbreak is attributable to NoV (80, 160). Outbreaks of NoV illness occur in a wide variety of different settings year-round but are particularly common and protracted in the winter months in closed settings (e.g., hospitals and nursing homes), where transmission is predominantly from person to person.

Determining the national burden of NoV outbreaks and comparing the burden between countries has been complicated by several factors. First, efforts to collect and test specimens from outbreaks of NoV-like illness tend to vary widely and are often less intensive than the response to outbreaks of suspected bacterial disease, where straightforward culture assays are routine. Second, diagnostic assays with different sensitivities and specificities for NoV, including electron microscopy, antigen detection assays, and PCR with diverse primers, are used in surveillance systems in different countries. Third, reporting systems have different inherent sensitivities for outbreaks in different settings and use disparate case definitions for outbreaks of viral gastroenteritis, including clinical criteria with no laboratory confirmation. One study in the United States showed a threefold variation between states in rates of reported food-borne outbreaks of any etiology. Moreover, only 11 of 50 states tested for NoV routinely, accounting for 75% of all reports of food-borne NoV outbreaks (175). Similarly, from 1995 to 2000, the number of viral gastroenteritis outbreaks reported in England and Wales ($n = 1,643$) was 40-fold that reported in France ($n = 43$), a country of similar population size (108).

The reported transmission modes and settings of outbreaks also vary from country to country: in the United States, 28% of NoV outbreaks confirmed at the Centers for Disease Control and Prevention (CDC) laboratory were classified as food-borne (12), compared to 17% of NoV outbreaks reported nationally in The Netherlands (90) and only 5% in England and Wales (107). It remains unclear if these are genuine disparities or an artifact of surveillance. In the United States, only food-borne outbreaks are formally reported and therefore are more likely to be tested and confirmed as NoV, whereas in the United Kingdom, high-profile hospital outbreaks in government-funded hospitals are nationally reportable, so proportionally more person-to-person outbreaks are likely to be documented. Some of the discordant data between countries may reflect genuine differences in epidemiology. Hospital outbreaks of NoV are relatively rarely reported in the United States despite a high frequency in nursing homes (12, 22). This may be explained by the fact that U.S. hospitals usually have one or two patients per room, whereas British hospitals frequently have several persons per room and occasionally have "Nightingale wards" with rows of beds on either side and shared bathroom facilities.

Although NoV has long been known to be transmitted by foods, the frequency and importance of food-borne transmission may have been overlooked, largely because of a lack of easy diagnosis. For example, of the 2,751 food-borne outbreaks reported to the CDC between 1993 and 1997, only 9 (0.3%) were confirmed as being due to NoV while 68% were of undetermined etiology (130). From 1990 to 2000 in the United States, the proportion of food-borne outbreaks attributable to NoV rose from <1% to 12% with the introduction of diagnostics to state public health laboratories (175). In 1998 in England and Wales, only 6% of food-borne outbreaks were reported as being associated with NoV (37), a low estimate compared to the United States; this may reflect the widespread use of relatively insensitive electron microscopy but may also point to a real difference in transmission in different countries.

To try to understand the role of NoV as a food-borne pathogen, several studies have used more active surveillance techniques involving RT-PCR assays to attempt to adjust for often limited collection and testing of stools from NoV-like outbreaks. In the United States, a study collected data from six states that tested most reported food-borne outbreaks for NoV to estimate that 50% of all food-borne outbreaks nationally may be attributable to NoV (175), consistent with an analysis of outbreaks in the state of Minnesota (29). Active surveillance in The Netherlands for 1 year estimated that NoV caused 31% of 49 food-borne outbreaks for which stools were collected, although a high proportion (37%) of these outbreaks were of underdetermined etiology despite comprehensive testing for viruses and bacteria (162). These findings are consistent with estimates derived in the United States using epidemiologic criteria specific for NoV-like illness, which have attributed 28% (160), 33% (55), and 38% (59) of all food-borne outbreaks to NoV.

Economic Burden

Food-borne disease imposes a very considerable economic burden on society through costs of measures to ensure safe food, workdays lost due to illness, health care costs, and economic impact of deaths. In the United States, the total cost of *Campylobacter* spp., *Listeria monocytogenes*, *Salmonella* spp. (nontyphoidal), *Escherichia coli* O157, and Shiga toxin-producing *E. coli* non-O157 in 2000 reached $6.9 billion (http://www.ers.usda.gov/briefing/FoodborneDisease/features.htm), but these pathogens account for only 25% of all food-borne disease (116). In The Netherlands, more than 80% of the costs associated with gastroenteritis were cost of time off work or the economic impact of deaths, highlighting the high nonmedical costs of food-borne illness (161). One British study estimated a total cost of $184 million nationally due to outbreaks of gastroenteritis (predominantly NoV) in 2002 and 2003 (109). There remain very few data on the economic impact of sporadic NoV disease, yet this work is important since, although NoV illness is unlikely to incur high costs due to chronic sequelae and deaths, the sheer frequency of these infections probably entails very substantial societal costs.

TREATMENT

No specific treatment exists for NoV illness. Therapy revolves around maintaining hydration and electrolyte balance until recovery. For patients with very frequent vomiting, the use of antiemetics has been considered. In experimentally infected adults, oral administration of bismuth subsalicylate reduced the severity of abdominal cramps and duration of gastrointestinal symptoms (149); however, the American Association of Pediatrics does not recommend the use of pharmacologic agents in children (4).

PREVENTION AND CONTROL

Control of NoV remains challenging since the viruses have a low infectious dose, are spread by multiple routes, contaminate widely and persist in the environment, and are more resistant to disinfectants than are many bacteria. Nonetheless, standard hygienic precautions such as regular hand washing and care in the disposal of vomitus and fecal material will help prevent NoV infection. Isolation of symptomatic persons into a separate airspace with dedicated toilet facilities, as far as is feasible, is important.

Efforts should also focus on exclusion of food handlers with gastroenteritis and possible furlough after diarrheal illness. In the United States, the Food and Drug Administration recently updated the Food Code to state that food workers should be excluded from the workplace for 24 h after resolution of any diarrheal illness (http://www.cfsan.fda.gov/~dms/fc05-toc.html). Although NoV may be shed for up to 3 weeks (145), even a short furlough from work may allow time for virus contamination on the body and clothes of a patient to be reduced. Once stools are formed, even though the feces may contain high viral loads, contamination of the environment and of the ill person with virus is less likely.

Disinfection

A lack of an infectivity assay for NoV has not allowed proper evaluation of the effectiveness of different disinfectants. One early study of volunteers who ingested Norwalk virus inocula with differing concentrations of chlorine suggested that NoVs may be resistant to up to 10 ppm of domestic bleach (85). However, the investigators used bovine fetal serum to dilute the inocula, and this may have inactivated much of the free chlorine. Moreover, waterborne outbreaks of NoV usually occur only during breakdown in chlorination of drinking water (94) or recreational water (141), which indicates that routine water chlorination (1 to 2 ppm) may be sufficient to inactivate NoVs in water (see chapter 8).

To evaluate disinfectants, efforts have turned to cultivable viruses from other genera within the family *Caliciviridae* as surrogate viruses because they grow well in cell culture and have similarities in capsid structure, genomic material, and genomic organization to NoV (see chapter 8). These viruses include feline calicivirus (FCV) and canine calicivirus (genus *Vesivirus*) and a cell culture-adapted strain of porcine enteric calicivirus belonging to the genus *Sapovirus*. FCV has been the most frequently used surrogate model for disinfection and survival studies of human NoVs. However, it may not be suitable as a surrogate since it is a respiratory virus and cannot survive at low pH, a necessary characteristic of enteric viruses that must survive passage through the stomach (35). Recently, murine NoV (MNV-1) was discovered in immunodeficient laboratory mice (81, 177), and this virus strain can be grown routinely in cell culture (17, 177). With the lack of a routine culture system for human NoV and its ability to survive at pH 2 for up to 2 h, MNV-1 may be a more appropriate surrogate model to measure the effectiveness of sanitizers (17) (Table 2). The U.S. Environmental Protection Agency has approved over 25 disinfectants for use against NoV largely on the basis of evidence from protocols using FCV (http://www.epa.gov/oppad001/chemregindex.htm).

Table 2 Comparison of the effects of different physiochemical parameters on murine NoV (MNV-1) and FCV infectivity[a]

Parameter	Exposure	Reduction of infectivity for different model systems	
		MNV-1	FCV
Acid stability	pH 2.0 for 30 min	$<1 \log_{10}$ PFU	$>4 \log_{10}$ PFU
Thermal stability	56°C	$D^b = 3.47$	$D = 6.75$
	63°C	$D = 0.435$	$D = 0.406$
	72°C	$D = 0.166$	$D = 0.118$
Environmental stability	4°C	$1 \log_{10}$	$1.5 \log_{10}$
(aqueous suspension)	25°C	$1.5 \log_{10}$	$4 \log_{10}$
Environmental stability	4°C	$2 \log_{10}$	$3 \log_{10}$
(surface)	25°C	$5.5 \log_{10}$	$5.0 \log_{10}$

[a]Derived from reference 17.
[b]D value signifies the decimal reduction time or the time (min) for 1-log reduction in titer at a given temperature.

Studies have demonstrated some efficacy of alcohol hand gels against FCV, but it remains unclear which concentration and type of alcohol would be the most effective (45, 78, 92). Current CDC advice is that, in the face of an outbreak, alcohol hand gels are of use in addition to more frequent hand washing (http://www.cdc.gov/ncidod/dhqp/id_norovirusFS.html).

Treatment of Food

Thorough cooking destroys NoVs, although some evidence points to resistance to temperatures up to 60°C for 30 min (32) and to steaming of shellfish (115). Studies of oysters have shown that NoVs bind specifically to carbohydrate receptors related to human blood group A antigens in the oyster intestine (99, 158). This implies that depuration (placing oysters in clean water for a period) is unlikely to decontaminate oysters. However, some work using murine NoV suggests that high-pressure treatment of oysters may inactivate NoV (86).

Vaccination

Obstacles to the development of a vaccine against NoV include the wide antigenic diversity of NoV strains, the lack in many individuals of either a heterotypic or homotypic protective immune response to infection, and the current lack of a cell culture or animal model for NoV. Nonetheless, the increasingly recognized burden of NoV illness has stimulated the exploration of several approaches to vaccine design. The first steps have shown immunogenicity of VLPs generated by baculovirus-infected insect cells when administered to mice (8, 124) and humans (155), as measured by serum immunoglobulin G and fecal immunoglobulin A levels. The intranasal route in mice may be more effective at eliciting a response to low doses of VLPs (52). Recombinant Norwalk virus particles have also proved immunogenic when expressed in transgenic potatoes and tomatoes, raising the intriguing possibility of an edible vaccine (154, 179). However, further steps in vaccine development depend on defining some correlate of protective immunity or a breakthrough in the search for an assay for neutralizing antibody.

CONCLUSION

Over the last decade, our understanding of NoV virology, pathogenesis, burden, and transmission has expanded greatly. However, challenges remain in several key areas (Table 3). First, further efforts are needed to develop cheap, easy-to-use, and accurate antigen detection assays and to shift diagnostics out of research and public laboratories and into clinical settings. This will allow larger and cheaper studies into disease burden in both developed and

Table 3 Knowledge gaps, challenges, and potential solutions

Knowledge gap	Hurdles	Solution
NoV as cause of severe gastroenteritis	No licensed rapid, cheap easy-to-use assay for NoV in clinical settings	Hospital-based studies with controls
	Perception that NoV infection causes only mild disease and so is not worth investigating	Improved investigation of severe illnesses in nursing homes
	Illness in vulnerable populations such as elderly persons in long-term care facilities is poorly documented	Development of sensitive, specific, and rapid commercial antigen detection assays
Frequency and severity of NoV disease in poor countries	RT-PCR assays often unavailable and too expensive	Hospital-based studies with controls, using existing platforms for research into enteric diseases
	Severe illness may not present at clinic, and deaths may go unreported	
Immune correlates of protection	Heterologous and homologous immunity not permanent in many individuals	Cell culture system for NoV to determine infectivity
	Genetic susceptibility factors confuse straightforward interpretation of antibody titers	Determine ABH genes/HBGA patterns in well-investigated outbreaks
Role of foods contaminated at source	Complex and varied food matrices make detection of NoV in food challenging	More routine stool collection and sequencing
	Dispersed outbreaks of common illness not easily linked	Investigation of possible sources of food contamination without assumption that it occurs at the point of service
		Real-time communication of sequence and epidemiologic data via the World Wide Web
		Development of improved laboratory protocols for detection of NoV in food
Effective environmental clean-up and disinfection	Persistent and resistant virus	Specific epidemiologic studies to compare field effectiveness of interventions
	Contamination widespread and may be unapparent	Culture system for NoV to assess infectivity
		Use of murine NoV as a surrogate virus

Table 3 *(continued)*

Knowledge gap	Hurdles	Solution
Reasons for predominance of GII.4 strains	Virulence factors for NoV unknown	Cell culture system for NoV to determine infectivity
		Identification of receptor(s) that are used by GII.4 and non-GII.4 strains
		Further investigation with volunteers or in outbreaks whether disease presentation or pathogenesis of certain strains is different

developing countries, as well as performance of more in-depth clinical investigations of unusual cases. Second, improved assays for testing of foods will help determine the frequency of contamination of food with NoV during processing. Third, the development of a culture system will allow for infectivity assays, a proper assessment of disinfectants, and a means of determining correlates of immunity to help efforts to develop a vaccine. Fourth, harmonization of assays to identify strains, along with rapid sharing of sequence data, will allow a more timely comparison of strains and discovery of links between cases and outbreaks, possibly due to a common source

Finally, in the last 10 years, as the profile of NoVs has increased along with the ability to diagnose infection, interest has spread from the domain of virologists to the wider public health and epidemiologic community. This has resulted in a much improved understanding of the disease burden and transmission of NoV. This dual approach to NoV research, virology, and public health epidemiology needs to continue if we are to develop better and more specific intervention strategies, including antiviral drugs and, hopefully, vaccines.

REFERENCES

1. **Adak, G. K., S. M. Long, and S. J. O'Brien.** 2002. Trends in indigenous foodborne disease and deaths, England and Wales: 1992 to 2000. *Gut* **51:**832–841.
2. **Agus, S. G., R. Dolin, R. G. Wyatt, A. J. Tousimis, and R. S. Northrup.** 1973. Acute infectious nonbacterial gastroenteritis: intestinal histopathology. Histologic and enzymatic alterations during illness produced by the Norwalk agent in man. *Ann. Intern. Med.* **79:**18–25.
3. **Akihara, S., T. G. Phan, T. A. Nguyen, G. Hansman, S. Okitsu, and H. Ushijima.** 2005. Existence of multiple outbreaks of viral gastroenteritis among infants in a day care center in Japan. *Arch. Virol.* **150:**2061–2075.

4. **American Academy of Pediatrics.** 1996. Practice parameter: the management of acute gastroenteritis in young children. *Pediatrics* **97**:424–435.

5. **Ando, T., S. S. Monroe, J. R. Gentsch, Q. Jin, D. C. Lewis, and R. I. Glass.** 1995. Detection and differentiation of antigenically distinct small round-structured viruses (Norwalk-like viruses) by reverse transcription-PCR and Southern hybridization. *J. Clin. Microbiol.* **33**:64–71.

6. **Armah, G. E., C. I. Gallimore, F. N. Binka, R. H. Asmah, J. Green, U. Ugoji, F. Anto, D. W. Brown, and J. J. Gray.** 2006. Characterisation of norovirus strains in rural Ghanaian children with acute diarrhoea. *J. Med. Virol.* **78**:1480–1485.

7. **Atmar, R. L., and M. K. Estes.** 2001. Diagnosis of noncultivatable gastroenteritis viruses, the human caliciviruses. *Clin. Microbiol. Rev.* **14**:15–37.

8. **Ball, J. M., M. E. Hardy, R. L. Atmar, M. E. Conner, and M. K. Estes.** 1998. Oral immunization with recombinant Norwalk virus-like particles induces a systemic and mucosal immune response in mice. *J. Virol.* **72**:1345–1353.

9. **Beller, M., A. Ellis, S. H. Lee, M. A. Drebot, S. A. Jenkerson, E. Funk, M. D. Sobsey, O. D. Simmons, S. S. Monroe, T. Ando, J. Noel, M. Petric, J. P. Middaugh, and J. S. Spika.** 1997. Outbreak of viral gastroenteritis due to a contaminated well. *JAMA* **278**:563–568.

10. **Bertolotti-Ciarlet, A., S. E. Crawford, A. M. Hutson, and M. K. Estes.** 2003. The 3′ end of Norwalk virus mRNA contains determinants that regulate the expression and stability of the viral capsid protein VP1: a novel function for the VP2 protein. *J. Virol.* **77**:11603–11615.

11. **Blacklow, N. R., G. Cukor, M. Bedigian, P. Echeverria, D. S. Greenberg, and J. S. Trier.** 1979. Immune response and prevalence of antibody to Norwalk enteritis virus as determined by radioimmunoassay. *J. Clin. Microbiol.* **10**:903–909.

12. **Blanton, L. H., S. M. Adams, R. S. Beard, G. Wei, S. N. Bulens, M. A. Widdowson, R. I. Glass, and S. S. Monroe.** 2006. Molecular and epidemiologic trends of caliciviruses associated with outbreaks of acute gastroenteritis in the United States, 2000–2004. *J. Infect. Dis.* **193**:413–421.

13. **Blutt, S. E., C. D. Kirkwood, V. Parreno, K. L. Warfield, M. Ciarlet, M. K. Estes, K. Bok, R. F. Bishop, and M. E. Conner.** 2003. Rotavirus antigenaemia and viraemia: a common event? *Lancet* **362**:1445–1449.

14. **Brugha, R., I. B. Vipond, M. R. Evans, Q. D. Sandifer, R. J. Roberts, R. L. Salmon, E. O. Caul, and A. K. Mukerjee.** 1999. A community outbreak of food-borne small round-structured virus gastroenteritis caused by a contaminated water supply. *Epidemiol. Infect.* **122**:145–154.

15. **Bull, R. A., E. T. Tu, C. J. McIver, W. D. Rawlinson, and P. A. White.** 2006. Emergence of a new norovirus genotype II.4 variant associated with global outbreaks of gastroenteritis. *J. Clin. Microbiol.* **44**:327–333.

16. **Burton-MacLeod, J., E. Kane, R. Glass, and T. Ando.** 2004. Evaluation and comparison of two commercial ELISA kits for detection of antigenically diverse human noroviruses in stool samples. *J. Clin. Microbiol.* **42**:2587–2595.

17. **Cannon, J. L., E. Papafragkou, G. W. Park, J. Osborne, L. A. Jaykus, and J. Vinjé.** 2006. Surrogates for the study of norovirus stability and inactivation in the environment: a comparison of murine norovirus and feline calicivirus. *J. Food Prot.* **69**:2761–2765.

18. Castilho, J. G., V. Munford, H. R. Resque, U. Fagundes-Neto, J. Vinjé, and M. L. Racz. 2006. Genetic diversity of norovirus among children with gastroenteritis in Sao Paulo State, Brazil. *J. Clin. Microbiol.* **44:**3947–3953.

19. Centers for Disease Control and Prevention. 2000. Foodborne outbreak of group A rotavirus gastroenteritis among college students—District of Columbia, March–April 2000. *Morb. Mortal. Wkly. Rep.* **49:**1131–1133.

20. Centers for Disease Control and Prevention. 2002. Outbreak of acute gastroenteritis associated with Norwalk-like viruses among British military personnel—Afghanistan, May 2002. *Morb. Mortal. Wkly. Rep.* **51:**477–479.

21. Centers for Disease Control and Prevention. 2003. Norovirus activity—United States, 2002. *Morb. Mortal. Wkly. Rep.* **52:**41–45.

22. Centers for Disease Control and Prevention. 2007. Norovirus activity—United States, 2006–2007. *Morb. Mortal. Wkly. Rep.* **56:**842–846.

23. Chan, M. C., J. J. Sung, R. K. Lam, P. K. Chan, N. L. Lee, R. W. Lai, and W. K. Leung. 2006. Fecal viral load and norovirus-associated gastroenteritis. *Emerg. Infect. Dis.* **12:** 1278–1280.

24. Cheetham, S., M. Souza, T. Meulia, S. Grimes, M. G. Han, and L. J. Saif. 2006. Pathogenesis of a genogroup II human norovirus in gnotobiotic pigs. *J. Virol.* **80:**10372–10381.

25. Chen, R., J. D. Neill, J. S. Noel, A. M. Hutson, R. I. Glass, M. K. Estes, and B. V. Prasad. 2004. Inter- and intragenus structural variations in caliciviruses and their functional implications. *J. Virol.* **78:**6469–6479.

26. Chiba, S., Y. Sakuma, R. Kogasaka, M. Akihara, K. Horino, T. Nakao, and S. Fukui. 1979. An outbreak of gastroenteritis associated with calicivirus in an infant home. *J. Med. Virol.* **4:**249–254.

27. Cotterelle, B., C. Drougard, J. Rolland, M. Becamel, M. Boudon, S. Pinede, O. Traore, K. Balay, P. Pothier, and E. Espie. 2005. Outbreak of norovirus infection associated with the consumption of frozen raspberries, France, March 2005. *Eurosurveillance* **10:** E050428 1.

28. de Bruin, E., E. Duizer, H. Vennema, and M. P. Koopmans. 2006. Diagnosis of Norovirus outbreaks by commercial ELISA or RT-PCR. *J. Virol. Methods* **137:**259–264.

29. Deneen, V. C., J. M. Hunt, C. R. Paule, O. I. James, R. G. Johnson, M. J. Raymond, and C. W. Hedberg. 2000. The impact of foodborne calicivirus disease: the Minnesota experience. *J. Infect. Dis.* **181:**S281–S283.

30. de Wit, M. A., M. P. Koopmans, L. M. Kortbeek, W. J. Wannet, J. Vinjé, F. van Leusden, A. I. Bartelds, and Y. T. van Duynhoven. 2001. Sensor, a population-based cohort study on gastroenteritis in the Netherlands: incidence and etiology. *Am. J. Epidemiol.* **154:**666–674.

31. Dimitriadis, A., and J. A. Marshall. 2005. Evaluation of a commercial enzyme immunoassay for detection of norovirus in outbreak specimens. *Eur. J. Clin. Microbiol. Infect. Dis.* **24:**615–618.

32. Dolin, R., N. R. Blacklow, H. DuPont, R. F. Buscho, R. G. Wyatt, J. A. Kasel, R. Hornick, and R. M. Chanock. 1972. Biological properties of Norwalk agent of acute infectious nonbacterial gastroenteritis. *Proc. Soc. Exp. Biol. Med.* **140:**578–583.

33. Dove, W., N. A. Cunliffe, J. S. Gondwe, R. L. Broadhead, M. E. Molyneux, O. Nakagomi, and C. A. Hart. 2005. Detection and characterization of human caliciviruses

in hospitalized children with acute gastroenteritis in Blantyre, Malawi. *J. Med. Virol.* 77:522–527.

34. Dowell, S. F., C. Groves, K. B. Kirkland, H. Cicirello, T. Ando, Q. Jin, J. R. Gentsch, S. S. Monroe, C. D. Humphrey, C. Slemp, D. M. Dwyer, R. A. Meriwether, and R. I. Glass. 1995. A multistate outbreak of oyster-associated gastroenteritis: implications for interstate tracing of contaminated shellfish. *J. Infect. Dis.* 171:1497–1503.

35. Duizer, E., P. Bijkerk, B. Rockx, A. De Groot, F. Twisk, and M. Koopmans. 2004. Inactivation of caliciviruses. *Appl. Environ. Microbiol.* 70:4538–4543.

36. Duizer, E., K. J. Schwab, F. H. Neill, R. L. Atmar, M. P. Koopmans, and M. K. Estes. 2004. Laboratory efforts to cultivate noroviruses. *J. Gen. Virol.* 85:79–87.

37. Evans, H. S., P. Madden, C. Douglas, G. K. Adak, S. J. O'Brien, T. Djuretic, P. G. Wall, and R. Stanwell-Smith. 1998. General outbreaks of infectious intestinal disease in England and Wales: 1995 and 1996. *Commun. Dis. Public Health* 1:165–171.

38. Evans, M. R., R. Meldrum, W. Lane, D. Gardner, C. D. Ribeiro, C. I. Gallimore, and D. Westmoreland. 2002. An outbreak of viral gastroenteritis following environmental contamination at a concert hall. *Epidemiol. Infect.* 129:355–360.

39. Falkenhorst, G., L. Krusell, M. Lisby, S. B. Madsen, B. Bottiger, and K. Molbak. 2005. Imported frozen raspberries cause a series of norovirus outbreaks in Denmark, 2005. *Eurosurveillance* 10:E050922 2.

40. Fankhauser, R. L., S. S. Monroe, J. S. Noel, C. D. Humphrey, J. S. Bresee, U. D. Parashar, T. Ando, and R. I. Glass. 2002. Epidemiologic and molecular trends of Norwalk-like viruses associated with outbreaks of gastroenteritis in the United States. *J. Infect. Dis.* 186:1–7.

41. Farkas, T., W. M. Zhong, Y. Jing, P. W. Huang, S. M. Espinosa, N. Martinez, A. L. Morrow, G. M. Ruiz-Palacios, L. K. Pickering, and X. Jiang. 2004. Genetic diversity among sapoviruses. *Arch. Virol.* 149:1309–1323.

42. Gabbay, Y. B., R. I. Glass, S. S. Monroe, C. Carcamo, M. K. Estes, J. D. Mascarenhas, and A. C. Linhares. 1994. Prevalence of antibodies to Norwalk virus among Amerindians in isolated Amazonian communities. *Am. J. Epidemiol.* 139:728–733.

43. Gallimore, C. I., C. Pipkin, H. Shrimpton, A. D. Green, Y. Pickford, C. McCartney, G. Sutherland, D. W. Brown, and J. J. Gray. 2005. Detection of multiple enteric virus strains within a foodborne outbreak of gastroenteritis: an indication of the source of contamination. *Epidemiol. Infect.* 133:41–47.

44. Gaulin, C. D., D. Ramsay, P. Cardinal, and M.-A. D'Halevyn. 1999. Epidemic of gastroenteritis of viral origin associated with eating imported raspberries. *Can. J. Public Health* 90:37–40.

45. Gehrke, C., J. Steinmann, and P. Goroncy-Bermes. 2004. Inactivation of feline calicivirus, a surrogate of norovirus (formerly Norwalk-like viruses), by different types of alcohol in vitro and in vivo. *J. Hosp. Infect.* 56:49–55.

46. Gerba, C. P., C. Wallis, and J. L. Melnick. 1975. Microbiological hazards of household toilets: droplet production and the fate of residual organisms. *Appl. Microbiol.* 30:229–237.

47. Gill, O. N., W. D. Cubitt, D. A. McSwiggan, B. M. Watney, and C. L. R. Bartlett. 1983. Epidemic of gastroenteritis caused by oysters contaminated with small round structured viruses. *Br. Med. J.* 287:1532–1534.

48. Graham, D. Y., X. Jiang, T. Tanaka, A. R. Opekun, H. P. Madore, and M. K. Estes. 1994. Norwalk virus infection of volunteers: new insights based on improved assays. *J. Infect. Dis.* **170:**34–43.

49. Gray, J. J., X. Jiang, P. Morgan-Capner, U. Desselberger, and M. K. Estes. 1993. Prevalence of antibodies to Norwalk virus in England: detection by enzyme-linked immunosorbent assay using baculovirus-expressed Norwalk virus capsid antigen. *J. Clin. Microbiol.* **31:**1022–1025.

50. Green, J., C. I. Gallimore, J. P. Norcott, D. Lewis, and D. W. G. Brown. 1995. Broadly reactive reverse transcriptase polymerase chain reaction for the diagnosis of SRSV-associated gastroenteritis. *J. Med. Virol.* **47:**392–398.

51. Green, K. Y. 2007. Human caliciviruses, p. 967–979. *In* D. M. Knipe and P. M. Howley (ed.), *Fields Virology*, 5th ed, vol. 1. Lippincott Williams & Wilkins, Philadelphia, PA.

52. Guerrero, R. A., J. M. Ball, S. S. Krater, S. E. Pacheco, J. D. Clements, and M. K. Estes. 2001. Recombinant Norwalk virus-like particles administered intranasally to mice induce systemic and mucosal (fecal and vaginal) immune responses. *J. Virol.* **75:**9713–9722.

53. Gurwith, M., W. Wenman, D. Gurwith, J. Brunton, S. Feltham, and H. Greenberg. 1983. Diarrhea among infants and young children in Canada: a longitudinal study in three northern communities. *J. Infect. Dis.* **147:**685–692.

54. Hall, G., M. D. Kirk, N. Becker, J. E. Gregory, L. Unicomb, G. Millard, R. Stafford, and K. Lalor. 2005. Estimating foodborne gastroenteritis, Australia. *Emerg. Infect. Dis.* **11:** 1257–1264.

55. Hall, J. A., J. S. Goulding, N. H. Bean, R. V. Tauxe, and C. W. Hedberg. 2001. Epidemiologic profiling: evaluating foodborne outbreaks for which no pathogen was isolated by routine laboratory testing: United States, 1982–9. *Epidemiol. Infect.* **127:**381–387.

56. Hansman, G. S., L. T. Doan, T. A. Kguyen, S. Okitsu, K. Katayama, S. Ogawa, K. Natori, N. Takeda, Y. Kato, O. Nishio, M. Noda, and H. Ushijima. 2004. Detection of norovirus and sapovirus infection among children with gastroenteritis in Ho Chi Minh City, Vietnam. *Arch. Virol.* **149:**1673–1688.

57. Hansman, G. S., K. Katayama, N. Maneekarn, S. Peerakome, P. Khamrin, S. Tonusin, S. Okitsu, O. Nishio, N. Takeda, and H. Ushijima. 2004. Genetic diversity of norovirus and sapovirus in hospitalized infants with sporadic cases of acute gastroenteritis in Chiang Mai, Thailand. *J. Clin. Microbiol.* **42:**1305–1307.

58. Harrington, P. R., J. Vinje, C. L. Moe, and R. S. Baric. 2004. Norovirus capture with histo-blood group antigens reveals novel virus-ligand interactions. *J. Virol.* **78:**3035–3045.

59. Hedberg, C. W., K. L. Palazzi-Churas, V. J. Radke, C. A. Selman, and R. V. Tauxe. 2007. The use of clinical profiles in the investigation of foodborne outbreaks in restaurants: United States, 1982–1997. *Epidemiol. Infect.* **2007:**1–8.

60. Hinkula, J., J. M. Ball, S. Lofgren, M. K. Estes, and L. Svensson. 1995. Antibody prevalence and immunoglobulin IgG subclass pattern to Norwalk virus in Sweden. *J. Med. Virol.* **47:**52–57.

61. Hjertqvist, M., A. Johansson, N. Svensson, P. E. Abom, C. Magnusson, M. Olsson, K. O. Hedlund, and Y. Andersson. 2006. Four outbreaks of norovirus gastroenteritis after consuming raspberries, Sweden, June–August 2006. *Eurosurveillance* **11:**E060907 1.

62. Hoebe, C. J., H. Vennema, A. M. de Roda Husman, and Y. T. van Duynhoven. 2004. Norovirus outbreak among primary schoolchildren who had played in a recreational water fountain. *J. Infect. Dis.* **189:**699–705.

63. Hohne, M., and E. Schreier. 2004. Detection and characterization of norovirus outbreaks in Germany: application of a one-tube RT-PCR using a fluorogenic real-time detection system. *J. Med. Virol.* **72:**312–319.

64. Huang, P., T. Farkas, W. Zhong, M. Tan, S. Thornton, A. L. Morrow, and X. Jiang. 2005. Norovirus and histo-blood group antigens: demonstration of a wide spectrum of strain specificities and classification of two major binding groups among multiple binding patterns. *J. Virol.* **79:**6714–6722.

65. Hutin, Y. J., V. Pool, E. H. Cramer, O. V. Nainan, J. Weth, I. T. Williams, S. T. Goldstein, K. F. Gensheimer, B. P. Bell, C. N. Shapiro, M. J. Alter, and H. S. Margolis for the National Hepatitis A Investigation Team. 1999. A multistate, foodborne outbreak of hepatitis A. *N. Engl. J. Med.* **340:**595–602.

66. Hutson, A. M., R. L. Atmar, D. Y. Graham, and M. Estes. 2002. Norwalk virus infection and disease is associated with ABO histo-blood group type. *J. Infect. Dis.* **185:**1335–1337.

67. Isakbaeva, E. T., M. A. Widdowson, R. S. Beard, S. N. Bulens, J. Mullins, S. S. Monroe, J. Bresee, P. Sassano, E. H. Cramer, and R. I. Glass. 2005. Norovirus transmission on cruise ship. *Emerg. Infect. Dis.* **11:**154–158.

68. Ito, S., S. Takeshita, A. Nezu, Y. Aihara, S. Usuku, Y. Noguchi, and S. Yokota. 2006. Norovirus-associated encephalopathy. *Pediatr. Infect. Dis. J.* **25:**651–652.

69. **Japan Ministry of Health and Welfare, National Institute of Infectious Diseases.** 2000. An outbreak of group A rotavirus infection among adults from eating meals prepared at a restaurant, April 2000—Shimane. *Infect. Agents Surveillance Rep.* **21:**145.

70. Jiang, X., D. Y. Graham, K. Wang, and M. K. Estes. 1990. Norwalk virus genome cloning and characterization. *Science* **250:**1580–1583.

71. Jiang, X., P. Huang, W. Zhong, A. L. Morrow, G. M. Ruiz-Palacios, and L. K. Pickering. 2004. Human milk contains elements that block binding of noroviruses to histo-blood group antigens in saliva. *Adv. Exp. Med. Biol.* **554:**447–450.

72. Jiang, X., J. Wang, D. Y. Graham, and M. K. Estes. 1992. Detection of Norwalk virus in stool by polymerase chain reaction. *J. Clin. Microbiol.* **30:**2529–2534.

73. Jiang, X., N. Wilton, W. M. Zhong, T. Farkas, P. W. Huang, E. Barrett, M. Guerrero, G. Ruiz-Palacios, K. Y. Green, J. Green, A. D. Hale, M. K. Estes, L. K. Pickering, and D. O. Matson. 2000. Diagnosis of human caliciviruses by use of enzyme immunoassays. *J. Infect. Dis.* **181** (Suppl. 2):S349–S359.

74. Johnson, P. C., J. J. Mathewson, H. L. DuPont, and H. B. Greenberg. 1990. Multiple-challenge study of host susceptibility to Norwalk gastroenteritis in US adults. *J. Infect. Dis.* **161:**18–21.

75. Jothikumar, N., J. A. Lowther, K. Henshilwood, D. N. Lees, V. R. Hill, and J. Vinjé. 2005. Rapid and sensitive detection of noroviruses by using TaqMan-based one-step reverse transcription-PCR assays and application to naturally contaminated shellfish samples. *Appl. Environ. Microbiol.* **71:**1870–1875.

76. Kageyama, T., S. Kojima, M. Shinohara, K. Uchida, S. Fukushi, F. B. Hoshino, N. Takeda, and K. Katayama. 2003. Broadly reactive and highly sensitive assay for Norwalk-like viruses based on real-time quantitative reverse transcription-PCR. *J. Clin. Microbiol.* **41:**1548–1557.

77. Kageyama, T., M. Shinohara, K. Uchida, S. Fukushi, F. B. Hoshino, S. Kojima, R. Takai, T. Oka, N. Takeda, and K. Katayama. 2004. Coexistence of multiple genotypes, includ-

ing newly identified genotypes, in outbreaks of gastroenteritis due to Norovirus in Japan. *J. Clin. Microbiol.* **42:**2988–2995.

78. Kampf, G., D. Grotheer, and J. Steinmann. 2005. Efficacy of three ethanol-based hand rubs against feline calicivirus, a surrogate virus for norovirus. *J. Hosp. Infect.* **60:**144–149.

79. Kapikian, A. Z., R. G. Wyatt, R. Dolin, T. S. Thornhill, A. R. Kalica, and R. M. Chanock. 1972. Visualization by immune electron microscopy of a 27-nm particle associated with acute infectious nonbacterial gastroenteritis. *J. Virol.* **10:**1075–1081.

80. Kaplan, J. E., R. Feldman, D. S. Campbell, C. Lookabaugh, and G. W. Gary. 1982. The frequency of a Norwalk-like pattern of illness in outbreaks of acute gastroenteritis. *Am. J. Public Health* **72:**1329–1332.

81. Karst, S. M., C. E. Wobus, M. Lay, J. Davidson, and H. W. Virgin IV. 2003. STAT1-dependent innate immunity to a Norwalk-like virus. *Science* **299:**1575–1578.

82. Katayama, K., H. Shirato-Horikoshi, S. Kojima, T. Kageyama, T. Oka, F. Hoshino, S. Fukushi, M. Shinohara, K. Uchida, Y. Suzuki, T. Gojobori, and N. Takeda. 2002. Phylogenetic analysis of the complete genome of 18 Norwalk-like viruses. *Virology* **299:**225–239.

83. Kaufman, S. S., N. K. Chatterjee, M. E. Fuschino, M. S. Magid, R. E. Gordon, D. L. Morse, B. C. Herold, N. S. LeLeiko, A. Tschernia, S. S. Florman, G. E. Gondolesi, and T. M. Fishbein. 2003. Calicivirus enteritis in an intestinal transplant recipient. *Am. J. Transplant.* **3:**764–768.

84. Kaufman, S. S., N. K. Chatterjee, M. E. Fuschino, D. L. Morse, R. A. Morotti, M. S. Magid, G. E. Gondolesi, S. S. Florman, and T. M. Fishbein. 2005. Characteristics of human calicivirus enteritis in intestinal transplant recipients. *J. Pediatr. Gastroenterol. Nutr.* **40:**328–333.

85. Keswick, B. H., T. K. Satterwhite, P. C. Johnson, H. L. DuPont, S. L. Secor, J. A. Bitsura, G. W. Gary, and J. C. Hoff. 1985. Inactivation of Norwalk virus in drinking water by chlorine. *Appl. Environ. Microbiol.* **50:**261–264.

86. Kingsley, D. H., D. R. Holliman, K. R. Calci, H. Chen, and G. J. Flick. 2007. Inactivation of a norovirus by high-pressure processing. *Appl. Environ. Microbiol.* **73:**581–585.

87. Kirkwood, C. D., R. Clark, N. Bogdanovic-Sakran, and R. F. Bishop. 2005. A 5-year study of the prevalence and genetic diversity of human caliciviruses associated with sporadic cases of acute gastroenteritis in young children admitted to hospital in Melbourne, Australia (1998–2002). *J. Med. Virol.* **77:**96–101.

88. Kojima, S., T. Kageyama, S. Fukushi, F. B. Hoshino, M. Shinohara, K. Uchida, K. Natori, N. Takeda, and K. Katayama. 2002. Genogroup-specific PCR primers for detection of Norwalk-like viruses. *J. Virol. Methods* **100:**107–114.

89. Koopmans, M., H. Vennema, H. Heersma, E. van Strien, Y. van Duynhoven, D. Brown, M. Reacher, and B. Lopman. 2003. Early identification of common-source foodborne virus outbreaks in Europe. *Emerg. Infect. Dis.* **9:**1136–1142.

90. Koopmans, M., J. Vinjé, M. de Wit, I. Leenen, and W. van der Poel. 2000. Molecular epidemiology of human enteric caliciviruses in The Netherlands. *J. Infect. Dis.* **181:**S262–S269.

91. Korsager, B., S. Hede, H. Boggild, B. E. Bottiger, and K. Molbak. 2005. Two outbreaks of norovirus infections associated with the consumption of imported frozen raspberries, Denmark, May–June 2005. *Eurosurveillance* **10:**E050623 1.

92. Kramer, A., A. S. Galabov, S. A. Sattar, L. Dohner, A. Pivert, C. Payan, M. H. Wolff, A. Yilmaz, and J. Steinmann. 2006. Virucidal activity of a new hand disinfectant with reduced ethanol content: comparison with other alcohol-based formulations. *J. Hosp. Infect.* **62**:98–106.

93. Kroneman, A., H. Vennema, J. Harris, G. Reuter, C. H. von Bonsdorff, K. O. Hedlund, K. Vainio, V. Jackson, P. Pothier, J. Koch, E. Schreier, B. E. Bottiger, and M. Koopmans. 2006. Increase in norovirus activity reported in Europe. *Eurosurveillance* **11**:E061214 1.

94. Kukkula, M., L. Maunula, E. Silvennoinen, and C. H. von Bonsdorff. 1999. Outbreak of viral gastroenteritis due to drinking water contaminated by Norwalk-like viruses. *J. Infect. Dis.* **180**:1771–1776.

95. Lambden, P. R., E. O. Caul, C. R. Ashley, and I. N. Clarke. 1993. Sequence and genome organization of a human small round-structured (Norwalk-like) virus. *Science* **259**:516–519.

96. Lau, C. S., D. A. Wong, L. K. Tong, J. Y. Lo, A. M. Ma, P. K. Cheng, and W. W. Lim. 2004. High rate and changing molecular epidemiology pattern of norovirus infections in sporadic cases and outbreaks of gastroenteritis in Hong Kong. *J. Med. Virol.* **73**:113–117.

97. Lee, N., M. C. Chan, B. Wong, K. W. Choi, W. Sin, G. Lui, P. K. Chan, R. W. Lai, C. S. Cockram, J. J. Sung, and W. K. Leung. High fecal viral concentration is associated with prolonged diarrhea in norovirus (genogroup II.4) gastroenteritis. *Emerg. Infect. Dis.*, in press.

98. Le Guyader, F., L. Haugarreau, L. Miossec, E. Dubois, and M. Pommepuy. 2000. Three-year study to assess human enteric viruses in shellfish. *Appl. Environ. Microbiol.* **66**:3241–3248.

99. Le Guyader, F., F. Loisy, R. L. Atmar, A. M. Hutson, M. K. Estes, N. Ruvoen-Clouet, M. Pommepuy, and J. Le Pendu. 2006. Norwalk virus-specific binding to oyster digestive tissues. *Emerg. Infect. Dis.* **12**:931–936.

100. Le Guyader, F. S., C. Mittelholzer, L. Haugarreau, K. O. Hedlund, R. Alsterlund, M. Pommepuy, and L. Svensson. 2004. Detection of noroviruses in raspberries associated with a gastroenteritis outbreak. *Int. J. Food Microbiol.* **97**:179–186.

101. Le Pendu, J. 2004. Histo-blood group antigen and human milk oligosaccharides: genetic polymorphism and risk of infectious diseases. *Adv. Exp. Med. Biol.* **554**:135–143.

102. Lew, J. F., J. Valdesuso, T. Vesikari, A. Z. Kapikian, X. Jiang, and M. K. Estes. 1994. Detection of Norwalk virus or Norwalk-like virus infections in Finnish infants and young children. *J. Infect. Dis.* **169**:1364–1367.

103. Lindesmith, L., C. Moe, S. Marionneau, N. Ruvoen, X. Jiang, L. Lindblad, P. Stewart, J. LePendu, and R. Baric. 2003. Human susceptibility and resistance to Norwalk virus infection. *Nat. Med.* **9**:548–553.

104. Liu, B. L., I. N. Clarke, E. O. Caul, and P. R. Lambden. 1995. Human enteric caliciviruses have a unique genome structure and are distinct from the Norwalk-like viruses. *Arch. Virol.* **140**:1345–1356.

105. Loisy, F., R. L. Atmar, P. Guillon, P. Le Cann, M. Pommepuy, and F. S. Le Guyader. 2005. Real-time RT-PCR for norovirus screening in shellfish. *J. Virol. Methods* **123**:1–7.

106. Lopman, B., H. Vennema, E. Kohli, P. Pothier, A. Sanchez, A. Negredo, J. Buesa, E. Schreier, M. Reacher, D. Brown, J. Gray, M. Iturriza, C. Gallimore, B. Bottiger, K. O. Hedlund, M. Torven, C. H. von Bonsdorff, L. Maunula, M. Poljsak-Prijatelj, J. Zimsek, G. Reuter, G. Szucs, B. Melegh, L. Svennson, Y. van Duijnhoven, and

M. Koopmans. 2004. Increase in viral gastroenteritis outbreaks in Europe and epidemic spread of new norovirus variant. *Lancet* **363**:682–688.

107. Lopman, B. A., G. K. Adak, M. H. Reacher, and D. W. Brown. 2003. Two epidemiologic patterns of norovirus outbreaks: surveillance in England and Wales, 1992–2000. *Emerg. Infect. Dis.* **9**:71–77.

108. Lopman, B. A., M. H. Reacher, Y. Van Duijnhoven, F. X. Hanon, D. Brown, and M. Koopmans. 2003. Viral gastroenteritis outbreaks in Europe, 1995–2000. *Emerg. Infect. Dis.* **9**:90–96.

109. Lopman, B. A., M. H. Reacher, I. B. Vipond, D. Hill, C. Perry, T. Halladay, D. W. Brown, W. J. Edmunds, and J. Sarangi. 2004. Epidemiology and cost of nosocomial gastroenteritis, Avon, England, 2002–2003. *Emerg. Infect. Dis.* **10**:1827–1834.

110. Lopman, B. A., M. H. Reacher, I. B. Vipond, J. Sarangi, and D. W. Brown. 2004. Clinical manifestation of norovirus gastroenteritis in health care settings. *Clin. Infect. Dis.* **39**:318–324.

111. Marks, P. J., I. B. Vipond, D. Carlisle, D. Deakin, R. E. Fey, and E. O. Caul. 2000. Evidence for airborne transmission of Norwalk-like virus (NLV) in a hotel restaurant. *Epidemiol. Infect.* **124**:481–487.

112. Marshall, J. A., M. E. Hellard, M. I. Sinclair, C. K. Fairley, B. J. Cox, M. G. Catton, H. Kelly, and P. J. Wright. 2003. Incidence and characteristics of endemic Norwalk-like virus-associated gastroenteritis. *J. Med. Virol.* **69**:568–578.

113. Marshall, J. K., M. Thabane, M. R. Borgaonkar, and C. James. 2007. Postinfectious irritable bowel syndrome after a food-borne outbreak of acute gastroenteritis attributed to a viral pathogen. *Clin. Gastroenterol. Hepatol.* **5**:457–460.

114. Mattner, F., D. Sohr, A. Heim, P. Gastmeier, H. Vennema, and M. Koopmans. 2006. Risk groups for clinical complications of norovirus infections: an outbreak investigation. *Clin. Microbiol. Infect.* **12**:69–74.

115. McDonnell, S., K. B. Kirkland, W. G. Hlady, C. Aristeguieta, R. S. Hopkins, S. S. Monroe, and R. I. Glass. 1997. Failure of cooking to prevent shellfish-associated viral gastroenteritis. *Arch. Intern. Med.* **157**:111–116.

116. Mead, P. S., L. Slutsker, V. Dietz, L. F. McCagi, J. S. Bresee, C. Shapiro, P. M. Griffin, and R. V. Tauxe. 1999. Food-related illness and death in the United States. *Emerg. Infect. Dis.* **5**:607–625.

117. Meeroff, J. C., D. S. Schreiber, J. S. B. Trier, and N. R. Blacklow. 1980. Abnormal gastric motor function in viral gastroenteritis. *Ann. Intern. Med.* **92**:370–373.

118. Moe, C. L., J. Gentsch, T. Ando, G. Grohmann, S. S. Monroe, X. Jiang, J. Wang, M. K. Estes, Y. Seto, C. Humphrey, S. Stine, and R. I. Glass. 1994. Application of PCR to detect Norwalk virus in fecal specimens from outbreaks of gastroenteritis. *J. Clin. Microbiol.* **32**:642–648.

119. Mohamed, N., S. Belak, K. O. Hedlund, and J. Blomberg. 2006. Experience from the development of a diagnostic single tube real-time PCR for human caliciviruses, Norovirus genogroups I and II. *J. Virol. Methods* **132**:69–76.

120. Morse, D. L., J. J. Guzewich, J. P. Hanrahan, R. Stricof, R. Shayegani, R. Deibel, J. C. Grabau, N. A. Nowak, J. E. Herrmann, G. Cukor, and N. R. Blacklow. 1986. Widespread outbreaks of clam- and oyster-associated gastroenteritis. Role of Norwalk virus. *N. Engl. J. Med.* **314**:678–681.

121. **Murata, T., N. Katsushima, K. Mizuta, Y. Muraki, S. Hongo, and Y. Matsuzaki.** 2007. Prolonged norovirus shedding in infants < or =6 months of age with gastroenteritis. *Pediatr. Infect. Dis. J.* **26:**46–49.

122. **Murphy, A. M., G. S. Grohmann, P. J. Christopher, W. A. Lopez, G. R. Davey, and R. H. Millsom.** 1979. An Australia-wide outbreak of gastroenteritis from oysters caused by Norwalk virus. *Med. J. Aust.* **2:**329–333.

123. **Nguyen, T. A., F. Yagyu, M. Okame, T. G. Phan, Q. D. Trinh, H. Yan, K. T. Hoang, A. T. Cao, P. Le Hoang, S. Okitsu, and H. Ushijima.** 2007. Diversity of viruses associated with acute gastroenteritis in children hospitalized with diarrhea in Ho Chi Minh City, Vietnam. *J. Med. Virol.* **79:**582–590.

124. **Nicollier-Jamot, B., A. Ogier, L. Piroth, P. Pothier, and E. Kohli.** 2004. Recombinant virus-like particles of a norovirus (genogroup II strain) administered intranasally and orally with mucosal adjuvants LT and LT(R192G) in BALB/c mice induce specific humoral and cellular Th1/Th2-like immune responses. *Vaccine* **22:**1079–1086.

125. **Noel, J. S., R. L. Fankhauser, T. Ando, S. S. Monroe, and R. I. Glass.** 1999. Identification of a distinct common strain of "Norwalk-like viruses" having a global distribution. *J. Infect. Dis.* **179:**1334–1344.

126. **Oh, D. Y., G. Gaedicke, and E. Schreier.** 2003. Viral agents of acute gastroenteritis in German children: prevalence and molecular diversity. *J. Med. Virol.* **71:**82–93.

127. **Oishi, I., K. Yamazaki, T. Kimoto, Y. Minekawa, E. Utagawa, S. Yamazaki, S. Inouye, G. S. Grohmann, S. S. Monroe, S. E. Stine, et al.** 1994. A large outbreak of acute gastroenteritis associated with astrovirus among students and teachers in Osaka, Japan. *J. Infect. Dis.* **170:**439–443.

128. **Okitsu-Negishi, S., M. Okame, Y. Shimizu, T. G. Phan, T. Tomaru, S. Kamijo, T. Sato, F. Yagyu, W. E. Muller, and H. Ushijima.** 2006. Detection of norovirus antigens from recombinant virus-like particles and stool samples by a commercial norovirus enzyme-linked immunosorbent assay kit. *J. Clin. Microbiol.* **44:**3784–3786.

129. **Oliver, S. L., E. Asobayire, A. M. Dastjerdi, and J. C. Bridger.** 2006. Genomic characterization of the unclassified bovine enteric virus Newbury agent-1 (Newbury1) endorses a new genus in the family Caliciviridae. *Virology* **350:**240–250.

130. **Olsen, S. J., L. C. MacKinnon, J. S. Goulding, N. H. Bean, and L. Slutsker.** 2000. Surveillance for foodborne-disease outbreaks-United States, 1993–1997. *CDC Surveill. Summ.* **49:**1–62.

131. **O'Ryan, M. L., N. Mamani, A. Gaggero, L. F. Avendano, S. Prieto, A. Pena, X. Jiang, and D. O. Matson.** 2000. Human caliciviruses are a significant pathogen of acute sproadic diarrhea in children of Santiago, Chile. *J. Infect. Dis.* **182:**1519–1522.

132. **O'Ryan, M. L., P. A. Vial, N. Mamani, X. Jiang, M. K. Estes, C. Ferrecio, H. Lakkis, and D. O. Matson.** 1998. Seroprevalance of Norwalk virus and Mexico virus in Chilean individuals: assessment of independent risk factors for antibody acquisition. *Clin. Infect. Dis.* **27:**789–795.

133. **Pang, X. L., J. K. Preiksaitis, and B. Lee.** 2005. Multiplex real time RT-PCR for the detection and quantitation of norovirus genogroups I and II in patients with acute gastroenteritis. *J. Clin. Virol.* **33:**168–171.

134. **Papaventsis, D. C., W. Dove, N. A. Cunliffe, O. Nakagomi, P. Combe, P. Grosjean, and C. A. Hart.** 2007. Norovirus infection in children with acute gastroenteritis, Madagascar, 2004–2005. *Emerg. Infect. Dis.* **13:**908–911.

135. Parashar, U. D., C. J. Gibson, J. S. Bresce, and R. I. Glass. 2006. Rotavirus and severe childhood diarrhea. *Emerg. Infect. Dis.* **12:**304–306.

136. Parashar, U. D., J.-F. Li, R. Cama, M. DeZalia, S. S. Monroe, D. N. Taylor, D. Figueroa, R. H. Gilman, and R. I. Glass. 2004. Human caliciviruses as a cause of severe gastroenteritis in Peruvian children. *J. Infect. Dis.* **190:**1088–1092.

137. Parrino, T. A., D. S. Schreiber, J. S. Trier, A. Z. Kapikian, and N. R. Blacklow. 1977. Clinical immunity in acute gastroenteritis caused by Norwalk agent. *N. Engl. J. Med.* **297:**86–89.

138. Peasey, A. E., G. M. Ruiz-Palacios, M. Quigley, W. Newsholme, J. Martinez, G. Rosales, X. Jiang, and U. J. Blumenthal. 2004. Seroepidemiology and risk factors for sporadic norovirus/Mexico strain. *J. Infect. Dis.* **189:**2027–2036.

139. Pelosi, E., P. R. Lambden, E. O. Caul, B. Liu, K. Dingle, Y. Deng, and I. N. Clarke. 1999. The seroepidemiology of genogroup 1 and genogroup 2 Norwalk-like viruses in Italy. *J. Med. Virol.* **58:**93–99.

140. Phan, T. G., S. Takanashi, K. Kaneshi, Y. Ueda, S. Nakaya, S. Nishimura, K. Sugita, T. Nishimura, A. Yamamoto, F. Yagyu, S. Okitsu, N. Maneekarn, and H. Ushijima. 2006. Detection and genetic characterization of norovirus strains circulating among infants and children with acute gastroenteritis in Japan during 2004–2005. *Clin. Lab.* **52:**519–525.

141. Podewils, L. J., L. Zanardi Blevins, M. Hagenbuch, D. Itani, A. Burns, C. Otto, L. Blanton, S. Adams, S. S. Monroe, M. J. Beach, and M. Widdowson. 2007. Outbreak of norovirus illness associated with a swimming pool. *Epidemiol. Infect.* **135:**827–833.

142. Ponka, A., L. Maunula, C. H. von Bonsdorff, and O. Lyytikainen. 1999. An outbreak of calicivirus associated with consumption of frozen raspberries. *Epidemiol. Infect.* **123:** 469–474.

143. Pusch, D., D. Y. Oh, S. Wolf, R. Dumke, U. Schroter-Bobsin, M. Hohne, I. Roske, and E. Schreier. 2005. Detection of enteric viruses and bacterial indicators in German environmental waters. *Arch. Virol.* **150:**929–947.

144. Richards, A. F., B. Lopman, A. Gunn, A. Curry, D. Ellis, H. Cotterill, S. Ratcliffe, M. Jenkins, H. Appleton, C. I. Gallimore, J. J. Gray, and D. W. Brown. 2003. Evaluation of a commercial ELISA for detecting Norwalk-like virus antigen in faeces. *J. Clin. Virol.* **26:**109–115.

145. Rockx, B., M. De Wit, H. Vennema, J. Vinjé, E. De Bruin, Y. Van Duynhoven, and M. Koopmans. 2002. Natural history of human calicivirus infection: a prospective cohort study. *Clin. Infect. Dis.* **35:**246–253.

146. Rockx, B. H., W. M. Bogers, J. L. Heeney, G. van Amerongen, and M. P. Koopmans. 2005. Experimental norovirus infections in non-human primates. *J. Med. Virol.* **75:**313–320.

147. Smiley, J. R., A. E. Hoet, M. Traven, H. Tsunemitsu, and L. J. Saif. 2003. Reverse transcription-PCR assays for detection of bovine enteric caliciviruses (BEC) and analysis of the genetic relationships among BEC and human caliciviruses. *J. Clin. Microbiol.* **41:**3089–3099.

148. Smit, T. K., P. Bos, I. Peenze, X. Jiang, M. K. Estes, and A. D. Steele. 1999. Seroepidemiological study of genogroup I and II calicivirus infections in South and southern Africa. *J. Med. Virol.* **59:**227–231.

149. Steinhoff, M. C., R. G. J. Douglas, and H. B. Greenberg. 1980. Bismuth subsalyctate therapy of viral gastroenteritis. *Gastroenterology* **78:**1495–1499.

150. Straub, T. M., K. Honer zu Bentrup, P. Orosz-Coghlan, A. Dohnalkova, B. K. Mayer, R. A. Bartholomew, C. O. Valdez, C. J. Bruckner-Lea, C. P. Gerba, M. Abbaszadegan, and C. A. Nickerson. 2007. In vitro cell culture infectivity assay for human noroviruses. *Emerg. Infect. Dis.* **13**:396–403.

151. Subekti, D. S., P. Tjaniadi, M. Lesmana, J. McArdle, D. Iskandriati, I. N. Budiarsa, P. Walujo, I. H. Suparto, I. Winoto, J. R. Campbell, K. R. Porter, D. Sajuthi, A. A. Ansari, and B. A. Oyofo. 2002. Experimental infection of *Macaca nemestrina* with a Toronto Norwalk-like virus of epidemic viral gastroenteritis. *J. Med. Virol.* **66**:400–406.

152. Sugieda, M., and S. Nakajima. 2002. Viruses detected in the caecum contents of healthy pigs representing a new genetic cluster in genogroup II of the genus "Norwalk-like viruses." *Virus Res.* **87**:165–172.

153. Swaminathan, B., T. J. Barrett, S. B. Hunter, and R. V. Tauxe. 2001. PulseNet: the molecular subtyping network for foodborne bacterial disease surveillance, United States. *Emerg. Infect. Dis.* **7**:382–389.

154. Tacket, C. O., H. S. Mason, G. Losonsky, M. K. Estes, M. M. Levine, and C. J. Arntzen. 2000. Human immune responses to a novel norwalk virus vaccine delivered in transgenic potatoes. *J. Infect. Dis.* **182**:302–305.

155. Tacket, C. O., M. B. Sztein, G. A. Losonsky, S. S. Wasserman, and M. K. Estes. 2003. Humoral, mucosal, and cellular immune responses to oral Norwalk virus-like particles in volunteers. *Clin. Immunol.* **108**:241–247.

156. Tan, M., P. Huang, J. Meller, W. Zhong, T. Farkas, and X. Jiang. 2003. Mutations within the P2 domain of norovirus capsid affect binding to human histo-blood group antigens: evidence for a binding pocket. *J. Virol.* **77**:12562–12571.

157. Tan, M., and X. Jiang. 2005. Norovirus and its histo-blood group antigen receptors: an answer to a historical puzzle. *Trends Microbiol.* **13**:285–293.

158. Tian, P., A. H. Bates, H. M. Jensen, and R. E. Mandrell. 2006. Norovirus binds to blood group A-like antigens in oyster gastrointestinal cells. *Lett. Appl. Microbiol.* **43**:645–651.

159. Trujillo, A. A., K. A. McCaustland, D. P. Zheng, L. A. Hadley, G. Vaughn, S. M. Adams, T. Ando, R. I. Glass, and S. S. Monroe. 2006. Use of TaqMan real-time reverse transcription-PCR for rapid detection, quantification, and typing of norovirus. *J. Clin. Microbiol.* **44**:1405–1412.

160. Turcios, R. M., M. A. Widdowson, A. C. Sulka, P. S. Mead, and R. I. Glass. 2006. Reevaluation of epidemiological criteria for identifying outbreaks of acute gastroenteritis due to norovirus: United States, 1998–2000. *Clin. Infect. Dis.* **42**:964–969.

161. van den Brandhof, W. E., G. A. De Wit, M. A. de Wit, and Y. T. van Duynhoven. 2004. Costs of gastroenteritis in The Netherlands. *Epidemiol. Infect.* **132**:211–221.

162. van Duynhoven, Y. T., C. M. de Jager, L. M. Kortbeek, H. Vennema, M. P. Koopmans, F. van Leusden, W. H. van der Poel, and M. J. van den Broek. 2005. A one-year intensified study of outbreaks of gastroenteritis in The Netherlands. *Epidemiol. Infect.* **133**:9–21.

163. Vernacchio, L., R. M. Vezina, A. A. Mitchell, S. M. Lesko, A. G. Plaut, and D. W. Acheson. 2006. Diarrhea in American infants and young children in the community setting: incidence, clinical presentation and microbiology. *Pediatr. Infect. Dis. J.* **25**:2–7.

164. Vinjé, J., J. Green, D. C. Lewis, C. I. Gallimore, D. W. Brown, and M. P. Koopmans. 2000. Genetic polymorphism across regions of the three open reading frames of "Norwalk-like viruses." *Arch. Virol.* **145**:223–241.

165. Vinjé, J., R. A. Hamidjaja, and M. D. Sobsey. 2004. Development and application of a capsid VP1 (region D) based reverse transcription PCR assay for genotyping of genogroup I and II noroviruses. *J. Virol. Methods* **116:**109–117.

166. Vinjé, J., and M. P. G. Koopmans. 1996. Molecular detection and epidemiology of small round structured viruses in outbreaks of gastroenteritis in the Netherlands. *J. Infect. Dis.* **174:**610–615.

167. Vinjé, J., H. Vennema, L. Maunula, C.-H. von Bonsdorff, M. Hoehne, E. Schreier, A. Richards, J. Green, D. Brown, S. S. Beard, S. S. Monroe, E. de Bruin, L. Svensson, and M. P. G. Koopmans. 2003. International collaborative study to compare reverse transcriptase PCR assays for detection and genotyping of noroviruses. *J. Clin. Microbiol.* **41:**1423–1433.

168. Wang, Q. H., M. G. Han, S. Cheetham, M. Souza, J. A. Funk, and L. J. Saif. 2005. Porcine noroviruses related to human noroviruses. *Emerg. Infect. Dis.* **11:**1874–1881.

169. Wheeler, C., T. M. Vogt, G. L. Armstrong, G. Vaughan, A. Weltman, O. V. Nainan, V. Dato, G. Xia, K. Waller, J. Amon, T. M. Lee, A. Highbaugh-Battle, C. Hembree, S. Evenson, M. A. Ruta, I. T. Williams, A. E. Fiore, and B. P. Bell. 2005. An outbreak of hepatitis A associated with green onions. *N. Engl. J. Med.* **353:**890–897.

170. Wheeler, J. G., D. Sethi, J. M. Cowden, P. G. Wall, L. C. Rodrigues, S. Tompkins, M. J. Hundson, and P. J. Roderick. 1999. Study of infectious intestinal disease in England: rates in the community, presenting to general practice, and reported to national surveillance. *Br. Med. J.* **318:**1046–1050.

171. Widdowson, M.-A., E. Cramer, L. Hadley, J. Bresee, R. S. Beard, S. Bulens, M. Charles, W. Chege, E. Isakbaeva, J. Wright, E. Mintz, J. Massey, R. I. Glass, and S. S. Monroe. 2004. Outbreaks of acute gastroenteritis on cruise ships and on land: identification of a predominant strain of norovirus (NV), United States 2002. *J. Infect. Dis.* **190:**27–36.

172. Widdowson, M. A., R. Glass, S. Monroe, R. S. Beard, J. W. Bateman, P. Lurie, and C. Johnson. 2005. Probable transmission of norovirus on an airplane. *JAMA* **293:**1859–1860.

173. Widdowson, M. A., M. I. Meltzer, X. Zhang, J. S. Bresee, U. D. Parashar, and R. I. Glass. 2007. Cost-effectiveness and potential impact of rotavirus vaccination in the United States. *Pediatrics* **119:**684–697.

174. Widdowson, M. A., B. Rockx, R. Schepp, W. H. van der Poel, J. Vinjé, Y. T. van Duynhoven, and M. P. Koopmans. 2005. Detection of serum antibodies to bovine norovirus in veterinarians and the general population in The Netherlands. *J. Med. Virol.* **76:**119–128.

175. Widdowson, M. A., A. Sulka, S. N. Bulens, R. S. Beard, S. S. Chaves, R. Hammond, E. D. Salehi, E. Swanson, J. Totaro, R. Woron, P. S. Mead, J. S. Bresee, S. S. Monroe, and R. I. Glass. 2005. Norovirus and foodborne disease, United States, 1991–2000. *Emerg. Infect. Dis.* **11:**95–102.

176. Widerlite, L., J. S. Trier, N. R. Blacklow, and D. S. Schreiber. 1975. Structure of the gastric mucosa in acute infectious nonbacterial gastroenteritis. *Gastroenterology* **68:**425–430.

177. Wobus, C. E., S. M. Karst, L. B. Thackray, K. O. Chang, S. V. Sosnovtsev, G. Belliot, A. Krug, J. M. Mackenzie, K. Y. Green, and H. W. Virgin. 2004. Replication of

Norovirus in cell culture reveals a tropism for dendritic cells and macrophages. *PLoS Biol.* 2:e432.

178. **Wolfaardt, M., M. B. Taylor, H. F. Booysen, L. Engelbrecht, W. O. K. Grabow, and X. Jiang.** 1997. Incidence of human calicivirus and rotavirus infection in patients with gastroenteritis in South Africa. *J. Med. Virol.* **51**:290–296.

179. **Zhang, X., N. A. Buehner, A. M. Hutson, M. K. Estes, and H. S. Mason.** 2006. Tomato is a highly effective vehicle for expression and oral immunization with Norwalk virus capsid protein. *Plant Biotechnol. J.* **4**:419–432.

180. **Zheng, D. P., T. Ando, R. L. Fankhauser, R. S. Beard, R. I. Glass, and S. S. Monroe.** 2006. Norovirus classification and proposed strain nomenclature. *Virology* **346**:312–323.

181. **Zintz, C., K. Bok, E. Parada, M. Barnes-Eley, T. Berke, M. A. Staat, P. Azimi, X. Jiang, and D. O. Matson.** 2005. Prevalence and genetic characterization of caliciviruses among children hospitalized for acute gastroenteritis in the United States. *Infect. Genet. Evol.* **5**:281–290.

Food-Borne Viruses: Progress and Challenges
Edited by Marion P. G. Koopmans, Dean O. Cliver, and Albert Bosch
© 2008 ASM Press, Washington, DC

Enterically Transmitted Hepatitis

3

Rakesh Aggarwal and Sita Naik

Viral hepatitis is a disease characterized by inflammation of the liver, which presents clinically as acute onset of jaundice preceded frequently by a short prodromal illness. It can be caused by infection with one of the several known hepatotropic viruses. These viruses differ widely in their morphology, genomic organization, taxonomic classification, and biological properties, as well as in their routes of spread and duration of persistence in the host (Table 1). All these viruses show a high degree of host-species specificity and tissue tropism for liver. In addition, several other systemic viral infections (due to Epstein-Barr virus, cytomegalovirus, yellow fever virus, etc.) may be associated with prominent liver injury and may therefore present with an illness resembling acute hepatitis. Two of the five well-characterized hepatitis viruses, namely hepatitis A virus (HAV) and hepatitis E virus (HEV), are transmitted by the enteral route and form the subject of this chapter.

HEPATITIS A

Virology and Molecular Biology

The HAV virions are small (27 to 32 nm in diameter), lack a lipid envelope, and have icosahedral symmetry. The viral genome consists of an approximately 7.5-kb, positive-sense, linear, single-stranded, polyadenylated RNA that has a small polypeptide (VPg) attached to its 5′ -end (8, 35). It contains a single open reading frame (ORF), which encodes a polyprotein of 252 kDa, which is processed to yield structural and nonstructural proteins (35). The amino-terminal end of the polyprotein encodes the viral structural proteins

RAKESH AGGARWAL, Department of Gastroenterology, Sanjay Gandhi Postgraduate Institute of Medical Sciences, Lucknow 226014, India. SITA NAIK, Department of Immunology, Sanjay Gandhi Postgraduate Institute of Medical Sciences, Lucknow 226014, India.

65

Table 1 Characteristics of major hepatitis viruses

Virus name	Viral genome	Taxonomic classification	Viral size and envelope	Route of transmission	Nature of infection and disease
Hepatitis A virus (HAV)	RNA	Family *Picornaviridae*, genus *Hepatovirus*	27–32 nm, nonenveloped	Fecal-oral	Acute hepatitis
Hepatitis B virus (HBV)	DNA	Family *Hepadnaviridae*	42 nm, enveloped	Parenteral	Acute and chronic hepatitis, liver cirrhosis and liver cancer
Hepatitis C virus (HCV)	RNA	Family *Flaviviridae*, genus *Hepacivirus*	Approximately 55 nm, lipid enveloped	Parenteral	Acute and chronic hepatitis, liver cirrhosis and liver cancer
Hepatitis D virus (HDV)	RNA	Unclassified	36 nm, enveloped (HBV envelope)	Parenteral	Acute and chronic hepatitis, liver cirrhosis, and liver cancer
Hepatitis E virus (HEV)	RNA	Family *Hepeviridae*, genus *Hepevirus*	27–34 nm, nonenveloped	Fecal-oral	Acute hepatitis

VP1, VP2, VP3, and the putative VP4, and the carboxy-terminal end encodes the nonstructural proteins associated with viral replication, including a protease and an RNA polymerase. A 734- to 740-nucleotide, 5' untranslated region helps in the translation process by folding into an internal ribosomal entry site (35).

Several cell culture lines support the growth of wild-type HAV, although at a relatively poor rate. However, HAV has been adapted to replicate well in many types of mammalian cells. This process of adaptation involves the development of mutations, several of which have been characterized in detail (5).

Based on genomic sequence data, various isolates of HAV have been classified into seven known genotypes (I to VII). Of these, three are simian in origin (genotypes IV, V, and VI); of the four known human HAV genotypes (genotypes I, II, III, and VII), genotypes I and III are the most prevalent (45). However, the various HAV genotypes, despite their <85% nucleotide identity, all belong to one serotype, allowing the use of a single vaccine. The genomic sequences of various HAV genotypes are, however, quite highly conserved in the 5' untranslated region, making it the most suitable genomic region for designing assays for detection and quantification of the virus in various clinical and other specimens.

Although HAV is hepatotropic, soon after entry into the host it appears to multiply in the intestine, from where it enters the circulation and reaches the liver. During HAV infection, the virus reaches the intestine in large quantities with bile and is shed in the feces. The virus is also shed in saliva in small quantities; this appears to have little clinical and epidemiological significance. HAV RNA has been detected in stool specimens collected 3 months or even longer after the onset of acute illness (61); however, infectivity of such fecal matter has not been shown. Prolonged excretion appears to be particularly common in young children.

Epidemiology

Epidemiological patterns

HAV infection is distributed worldwide. However, its endemicity in different parts of the world is closely linked to the level of socioeconomic development. In developing countries with poor sanitary and living conditions, such as those in Africa and parts of Asia and South America, transmission rates of HAV infection are high and most infections occur in early childhood. HAV infection is known to confer strong protection against reinfection; hence, infection among adults is uncommon in these regions. In contrast, in developed countries with lower HAV endemicity, transmission rates are much lower.

Age at infection is a major determinant of the clinical expression of HAV infection, with the likelihood of disease increasing with increasing age. Thus, most infections in children younger than 6 years are either asymptomatic or associated with mild and nonspecific symptoms, whereas those in older children and adults are usually symptomatic and associated with icteric disease (18, 31). Also, the risk of complications and the case fatality rate increase with increasing age (59).

Based on the above factors, four major epidemiological patterns of HAV infection have been described (Fig. 1). In countries where HAV infection is highly endemic (such as India), high transmission rates lead to near universal infection in early childhood, with infrequent clinical cases. In areas of intermediate endemicity (e.g., China), the virus circulates at a fairly high rate in a population which has a substantial proportion of susceptible older children and young adults, in whom infection is frequently symptomatic; thus, these areas show a high incidence of clinical disease. These cases occur in the form of large outbreaks which may be related to person-to-person transmission or may be water or food borne. In areas of low endemicity (e.g., the United States), disease rates are low, even though most of the older children,

Figure 1 Epidemiological patterns of HAV infection.

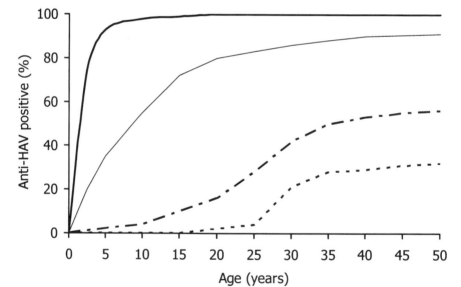

adolescents, and young adults are susceptible to HAV infection, because of fewer opportunities for exposure; the cases of hepatitis A in these regions arise from person-to-person spread or from common-source, food-borne outbreaks. In areas of very low endemicity (e.g., Sweden), there is virtually no indigenous transmission of disease and the infrequent cases are mostly related to travel to regions where the infection is endemic.

Routes of transmission

HAV is transmitted via the fecal-oral route. The most common route of transmission is through close contact with an infected person, usually in a household or school setting. Even in developed countries with high levels of personal hygiene, the rates of transmission of HAV to household contacts of patients with hepatitis A are very high, indicating the importance of this route (50). A large proportion of hepatitis A cases do not have an identifiable source of infection; such cases probably arise from personal contact with an unidentified, frequently asymptomatic, person shedding HAV.

Contaminated food and water are less common sources of transmission of HAV than is personal contact. However, such contamination may lead to outbreaks of hepatitis A, particularly in regions of low or intermediate endemicity. Food-borne transmission is discussed in more detail in a separate section below.

HAV infection can be transmitted by transfusion of blood or blood products; however, such cases are uncommon. Outbreaks of HAV infection have been reported among injection-drug users (23). Transmission in these outbreaks is believed to be multifactorial, being related to person-to-person spread via close contact, poor personal hygiene among drug users, and percutaneous spread via shared drugs, needles, and other equipment used to prepare the drugs prior to injection. Cases and outbreaks related to sexual transmission of HAV, especially those among men having sex with men, have also been reported.

Diagnosis, Serology, Protective Antibodies, and Vaccine Status

HAV infection induces the development of specific immunoglobulin M (IgM) and IgG antibodies. The IgM antibodies usually appear before the onset of clinical disease, are directed primarily against the viral capsid protein, and persist for 4 to 6 months. Their presence serves as a useful marker for the diagnosis of acute HAV infection. Several laboratory methods have been used to detect IgM anti-HAV antibodies; these include radioimmunoassays, enzyme-linked immunosorbent assays (EIA), and immunoblotting. Of these, EIAs are the most commonly used.

The IgG anti-HAV antibodies, in contrast, persist for several years. The presence of detectable IgG antibodies in most of the elderly individuals in populations where circulation of HAV has been absent for several decades suggests that they persist for life. The anti-HAV antibodies are neutralizing in nature and provide complete protection against reinfection.

Immunization with HAV vaccines is associated with the development of IgG antibodies against the viral structural proteins. However, the titer of vaccine-induced antibodies is lower than that following natural infection, and the commercially available anti-HAV tests may not be sensitive enough to detect these antibodies in a significant proportion of vaccinees, particularly if testing is done several years after immunization (32). Some of the vaccinated persons also develop a transient IgM anti-HAV response.

Thermal and Environmental Stability of HAV

HAV is stable in the environment for at least 1 month (38), especially when associated with organic matter. It is more resistant to heating and chlorine inactivation than is poliovirus. It has been shown to survive on human hands and fomites for long periods (37). These characteristics facilitate its transmission by person-to-person contact and through contaminated food and water. HAV can remain infectious in frozen foods for periods exceeding a year (22).

HAV can be inactivated by heating foods to temperatures above 85°C for 1 min. Surfaces contaminated with HAV can be disinfected by contact with a 1:100 dilution of sodium hypochlorite for 1 min (42).

Evidence of Food-Borne Transmission

Several food-borne outbreaks of hepatitis A have been reported (see chapter 1). Contamination of several foods and food products, including seafoods, fruits, and vegetables, has been implicated in these outbreaks. Several types of seafood, including clams, mussels, and oysters, have been implicated in the transmission of HAV (9, 12). In recent years, outbreaks of hepatitis A associated with farm products such as strawberries and green onions have been recorded (22, 58). Outbreaks have also followed the consumption of contaminated ice-slush beverages and salad food items (4, 34). Some of these outbreaks have been very large, as in the case of an outbreak in China which caused illness in nearly 300,000 persons and is thought to be related to ingestion of raw clams (20).

The evidence linking the ingestion of contaminated foods with the occurrence of clinical cases of hepatitis A is multipronged and quite strong. It includes epidemiological evidence supporting common-source food ingestion, evidence of HAV infection in suspected food handlers, molecular data showing identical viral isolates from affected patients residing in geographically diverse locations but ingesting the same food, and close genetic relationship between viral isolates from the patients and the presumed source of contamination.

Mechanisms of Entry and Concentration of Virus in Foods

Humans are the only source of HAV. After excretion in human feces, the virus can contaminate food items though several possible routes. Fruits and vegetables are most often contaminated through contact with contaminated water or treated sewage, either during irrigation or harvesting at farms, or during processing and distribution. Food products may also become contaminated through contact with an infected food handler or from contaminated surfaces during their preparation.

Bivalves (mollusks) concentrate HAV inside themselves by a processs akin to filtration. These organisms take up large quantities of water, which passes through while the viral particles contained in the water are trapped. This leads to virus concentrations in the bivalves that are severalfold higher than those in the water. This bioaccumulation is further helped by binding of viral particles to mucopolysaccharides in shellfish mucus. Whether plants have a similar mechanism for concentrating HAV and other viruses remains unclear.

Prevention of Food-Borne Transmission

The most reliable methods for preventing food-borne hepatitis A include (i) adequate cooking of food and (ii) adhering to hygiene precautions during preparation and handling of food. Hygiene precautions are particularly important for foods which are either eaten uncooked or handled but not adequately reheated between being cooked and being consumed. Some persons prefer to eat shellfish either raw or after minimal cooking, because of their taste preferences; this practice needs to be discouraged through health education and regulation of restaurants serving such food. At the farm, contamination of food can be reduced by ensuring that clean water is used for irrigation and for rinsing of vegetables and other produce. Providing adequate sanitary facilities for workers and ensuring that only healthy workers handle food products at farms, food-processing centers, and restaurants are other steps that can help break the chain of food-borne transmission of hepatitis A.

Depuration is a commercial process whereby live shellfish stock are kept in tanks of clean seawater for long periods extending over several days in an attempt to rid them of viruses. However, this technique is only partially successful and appears to be inadequate by itself (16). Rinsing of vegetable produce with chlorinated water reduces but does not eliminate contamination by

HAV. UV irradiation can also decrease the virus concentration in shellfish (44); however, the dose required acts adversely on the taste, appearance, and shelf life of the product, making this a nonviable option. In one study, even the use of an advanced closed-circuit depuration system which also used both ozone and UV light to disinfect the water failed to eliminate HAV from contaminated mussels (11).

Treatment with high hydrostatic pressure to inactivate HAV in food has been tried. In one study, application of 350, 375, and 400 MPa of pressure for 1 min at 8.7 to 10.3°C led to >1-, >2- and >3-$\log_{10}$ reduction, respectively, in the viral content of live oysters that had each been contaminated in vitro with $>10^5$ PFU of HAV (7).

Since the methods used to inactivate HAV are only partially successful, cooking to high temperatures and attention to hygiene precautions continue to be the primary preventive measures.

Detection Methods for Contamination of Food

Traditionally, tests for the presence of coliform bacteria and *Escherichia coli* have been used to assess the sanitary quality of seafood. These tests prevent food-borne infections by fecal bacteria; although the tests serve as surrogate markers for feces-derived viruses, their results do not always correlate with the presence of viruses such as HAV.

Detection of HAV using immune electron microscopy or cell culture techniques and detection of its antigens using sensitive immunological techniques such as radioimmunoassay and EIA have been tried. However, the concentration of HAV in food and environmental specimens is much lower than the lower detection limit of these techniques, precluding their use for surveillance of food products.

The detection of HAV in food material and environmental specimens is therefore based on specific methods such as reverse transcription-PCR (RT-PCR) for detecting small quantities of viral nucleic acids; these methods involve amplification of viral genomic sequences, making them highly sensitive (see chapter 7). Attempts have been made to modify these assays to counter the effect of the presence in foods of several substances that inhibit nucleic acid amplification. For instance, immunomagnetic pretreatment of specimens using magnetic beads coated with anti-HAV antibodies improves the sensitivity of the PCR assay for HAV RNA in artificially spiked food rinses by 10 to 20-fold. Further, this pretreatment procedure combined with real-time RT-PCR could be carried out within 6 h, making it particularly suitable for routine surveillance of foods (47). However, the tests for detection of HAV have not yet reached a level where they can be used to test food at farms or in food-processing centers.

Molecular epidemiologic techniques such as HAV genotyping and nucleic acid sequencing can be applied to food specimens. Demonstration of identical or closely related HAV genomic sequences in a suspected food item and in clinical specimens from patients may help establish food-borne transmission. In addition, these findings may help trace the geographical origin of the contaminating virus, based on prior knowledge of the virus isolates prevalent in various regions.

Vaccination for Postexposure Prophylaxis
Both active and passive prophylaxis have been used to protect against hepatitis A. Passive prophylaxis can be done using immune serum globulin administered at a dose of 0.02 ml/kg of body weight, given by the intramuscular route, within 2 weeks of exposure; this technique has a high protective efficacy. However, with the availability of effective hepatitis A vaccines, which have the additional advantage of providing long-term protection, the use of immune globulin has declined considerably.

Two different hepatitis A vaccines are commercially available (24, 57). Both of these contain HAV grown in cell culture, which has been purified and formalin inactivated, along with alum as an adjuvant. The two vaccines differ in the HAV strains used. For each vaccine, two intramuscular doses in the deltoid region, separated by 6 to 12 months, are advised. Both the vaccines are highly immunogenic, have good protective efficacy against all HAV genotypes, and are very safe. These vaccines are useful not only for preexposure prophylaxis but also for postexposure prophylaxis when administered in the early part of the incubation period. Hence, they have been extensively used to control food-borne outbreaks and to reduce the risk of infection among close contacts of identified patients with hepatitis A.

HEPATITIS E
Virology, Molecular Biology, and Genotypes
The HEV virions are 32 to 34 nm in diameter, nonenveloped, and icosahedral in symmetry (6). The HEV genome is a single-stranded, positive-sense, and polyadenylated RNA genome of approximately 7,200 bases, which carries a 7-methylguanosine (m^7G) nucleotide cap. It contains three partially overlapping ORFs, which are flanked by a 5′ untranslated region and a 3′ poly(A) tail (53). ORF1 codes for a polyprotein, which is thought to be processed to yield viral nonstructural proteins (RNA-dependent RNA polymerase, helicase, and methyl transferase) that may play a role in viral replication. ORF2 encodes the major structural (capsid) protein of the virus. ORF3

encodes a small phosphoprotein that appears to be involved in the assembly of newly formed virions and in intracellular signal transduction.

It has not been possible to grow HEV efficiently and reliably in a cell culture system; this failure has hampered the study of replication and other biological aspects of this virus. However, in recent years, infectious cDNA clones of HEV and a subgenomic replicon system have been developed (15). These developments, along with recent success at culturing HEV in a hepatocarcinoma cell line (51) and in monolayers of HepG2/C3A cells (14), should permit clarification of some of these issues in the near future.

Based on the phylogenetic relationship of their genomic sequences, HEV isolates have been classified into at least four genotypes (46). Genotype 1 includes several Asian isolates, which have a nucleotide sequence homology of 92 to 99% (amino acid sequence homology of 95 to 99%) to each other. Genotype 2 includes a Mexican strain, which has only 75% nucleotide sequence homology (86% amino acid homology) to the genotype 1 isolates, and a few recently described African isolates. Genotype 3 comprises isolates from sporadic cases in the United States, an area where HEV is not endemic; these isolates are 92% identical to each other but only 73.5 to 74.5% identical to those belonging to genotypes 1 and 2. It also includes swine HEV isolates from the United States, Europe, and Japan, as well as human HEV isolates from these regions of nonendemicity. Some human HEV isolates from China and Taiwan and swine HEV isolates from India belong to a separate genotype, genotype 4. Genotypes 5 to 8 have been proposed for solitary isolates from other regions of nonendemicity such as Italy, Greece, and Spain. However, as with HAV, all HEV genotypes share serological epitopes and belong to one serotype.

Swine HEV, an HEV-like virus, was first isolated in 1997 from pigs in the United States (40). Since then, similar reports have appeared from several countries, including those where human HEV infection is uncommon and those where such infection is endemic. Natural infection with swine HEV in pigs leads to transient viremia and development of antibodies that cross-react with the capsid protein of the human HEV. Swine HEV has 90 to 92% and 79 to 83% identity at the amino acid level in the ORF2 and ORF3 regions of the Burmese and Mexican isolates of human HEV, respectively, but is more closely related to the U.S. isolates of human HEV. Also, genetic sequences of swine HEV strains in Taiwan and Spain more closely resemble those of human HEV strains from these regions than they resemble swine and human HEV isolates from other parts of the world. Genotype 3 human and swine HEV are capable of cross-species infection; thus, swine HEV can infect nonhuman primates and the US-2 strain of human HEV can infect pigs (19, 39). These observations have suggested the possibility that HEV infection is a zoonosis. However, swine HEV isolates from India, an HEV-endemic region, are genetically different from human HEV isolates from patients from the same geographical region; the isolates belong to genotypes 4 and 1, respectively (3).

An HEV-like virus has been recovered from chickens with hepatomegaly-splenomegaly syndrome. This virus appears to be enzootic in chickens, with frequent subclinical spread. Its genome is about 600 bases shorter than those of human and swine HEV isolates, and has nearly 50% nucleotide homology to human and swine HEVs (21). On phylogenetic analysis, the avian HEV forms a branch distinct from the mammalian (human and swine) HEV isolates. Of its several B-cell antigenic domains, some are shared with mammalian HEVs, whereas others are unique. Currently, avian HEV is not thought to play any role in human disease.

Epidemiology

HEV infection is endemic in many parts of Asia, Africa, the Middle East, and Central America. In these areas, cases of hepatitis E occur in the form of epidemics as well as sporadic cases.

HEV infection is transmitted primarily by the fecal-oral route. Most of the well-recorded outbreaks of hepatitis E have been caused by fecal contamination of drinking-water supplies (26, 41, 55). The outbreaks often follow periods of heavy rain leading to flooding. Some outbreaks have, however, been reported to occur during hot summer months, when reduced water flow rates in rivers and streams lead to increased concentrations of fecal contaminants. In several HEV-endemic regions, hepatitis E epidemics occur with a periodicity of 5 to 10 years; reasons for this periodicity remain unclear.

The first major recorded outbreak of waterborne hepatitis was observed in Delhi, India, in 1955 to 1956 (55). This outbreak, with nearly 29,000 cases of icteric hepatitis, was initially thought to have been caused by HAV and was serologically documented as non-A, non-B hepatitis only several years later. Subsequently, large epidemics of hepatitis E have been reported from the Indian subcontinent, including Kashmir (26), Kanpur (with nearly 79,000 cases) (41), and Nepal. Many hepatitis E outbreaks have involved several thousand cases; the largest reported to date, involving more than 100,000 cases, occurred in northwestern China between 1986 and 1988 (62). Small outbreaks of hepatitis E were reported from two villages located south of Mexico City several years ago (54); however, no further outbreaks have been reported in this region since then. In recent years, large hepatitis E outbreaks have been reported from Africa (43).

Person-to-person transmission of HEV among family members of patients with epidemic and sporadic hepatitis E has been uncommon (1, 49).

Reported secondary attack rates for households with hepatitis E-infected individuals have ranged from 0.7 to 2.2% compared with 50 to 75% among susceptible contacts in households with hepatitis A-infected persons. Vertical transmission of infection from HEV-infected pregnant mothers to their infants has been shown to occur (28). Although evidence of transmission of HEV infection through blood transfusions has been reported in a study from an HEV-endemic region (29), the contribution of this route to the total number of cases of hepatitis E remains unclear.

In regions with hepatitis E outbreaks, HEV infection also accounts for a substantial proportion of cases of acute sporadic hepatitis in both children and adults. In these HEV-endemic countries, nearly 5% of children younger than 10 years and 10 to 40% of adults show the presence of detectable anti-HEV antibodies in their sera. These rates appear to be somewhat low in view of the frequent occurrence of outbreaks and nearly ubiquitous opportunities for infection, raising the possibility that these antibodies might disappear with time. Even in several countries where outbreaks have not been reported, such as Hong Kong, Turkey, and Egypt, the disease appears to be endemic since a significant proportion of sporadic hepatitis cases in these countries are related to HEV infection. Recent data suggest that a proportion of healthy persons in HEV-endemic regions have detectable IgM anti-HEV and HEV RNA in their blood; the exact significance of these findings is not yet clear.

Clinical Features and Outcome

The clinical features of hepatitis E are no different from those of other forms of acute viral hepatitis. Most patients have a self-limiting course and their condition improves spontaneously over a few weeks. However, a minority of patients develop acute liver failure, which may be fatal. In a significant proportion of those infected, the infection may pass entirely unrecognized; such a subclinical or anicteric course appears to be particularly common among children. During outbreaks, disease attack rates are much higher among pregnant women than among men and nonpregnant women (26). In addition, pregnant women are more likely to develop acute liver failure. HEV infection does not lead to chronic hepatitis.

The mortality rate due to hepatitis E in large outbreaks has been relatively low. In a large outbreak affecting nearly 79,000 persons, the mortality rate was only 0.07% (41). Hospital-based data have shown higher death rates of up to 4%; however, these rates probably represent a selection bias. The mortality rates are particularly high among pregnant women with hepatitis E, possibly as high as 15 to 25% (26).

In recent years, HEV superinfection in persons with preexisting though often asymptomatic chronic liver disease has been shown to lead to the clinical syndrome of decompensated or acute-on-chronic liver disease (30).

Diagnosis, Serology, Protective Antibodies, and Vaccine Status

HEV infection is associated with development of specific IgM antibodies, which appear by the time of onset of clinical illness and persist for a few months (10). Detection of these antibodies is used to diagnose acute HEV infection.

The IgG anti-HEV antibodies appear simultaneously with or shortly after the IgM response. However, persistence of these antibodies is uncertain. In one study, nearly half of those who had been infected during an outbreak 14 years earlier tested positive for anti-HEV IgG (27). This was taken as evidence of persistence of anti-HEV antibodies for long periods; however, the possibility of reinfection with HEV during the long intervening period could not be excluded.

Although several assay formats have been developed for the detection of anti-HEV antibodies of various subclasses, EIA remains the most widely used format. Recently, a rapid immunochromatographic test for anti-HEV IgM was developed and has shown promise for use at the point of care.

Thermal and Environmental Stability of HEV

Data on the thermal and environmental stability of HEV are quite limited. In a recent study, heating of HEV-containing fecal suspensions to between 45 and 70°C led to a reduction of residual infectivity of the virus (13). These studies suggested that, despite being less heat stable than HAV, at least some HEV virions would be expected to survive the temperatures reached inside rare-cooked meat.

The persistence of infectivity of HEV under environmental conditions has not been studied. However, the occurrence of large outbreaks following contamination of water sources with sewage suggests that the virus remains viable for a sufficient duration to survive the several steps and long time gap between excretion and entry into a susceptible host.

The effect of chlorine, a widely used agent for water purification, in the inactivation of HEV has not been directly studied. However, a large outbreak of hepatitis E in Kanpur, India, was temporally related to the failure of chlorination of the water supply (41), suggesting indirectly that chlorination plays a role in the inactivation of HEV.

Evidence of Food-Borne Transmission

Evidence supporting transmission of HEV infection through food other than water is much more limited than that for HAV. This is possibly because the long incubation period of 2 to 10 weeks makes it difficult to identify a common-source food exposure in patients with hepatitis E. Also, a significant proportion of those infected may have a subclinical infection; thus, even if several persons are infected from a common food source, only a minority of them may develop clinically identifiable illness. Furthermore, in HEV-endemic regions, several large, well-characterized waterborne outbreaks of HEV infection have been reported; since opportunities for water contamination are ubiquitous, there may be a tendency to characterize even minor outbreaks as waterborne, even though some of these may be related to shared food. A large proportion of sporadic cases of hepatitis E that are common in HEV-endemic countries may in fact represent cases of unrecognized food-borne transmission.

The relative lack of evidence supporting food-borne transmission of HEV in HEV-endemic regions may also be related to the fact that extensive epidemiological investigations for food-borne infections are undertaken only infrequently in these areas. Also, even for hepatitis A, well-defined food-borne outbreaks have been recorded primarily in developed countries with low background rates of disease and facilities for extensive epidemiological and laboratory investigations.

Most reports on food-borne transmission of HEV have been from Japan. An early report described two patients who had eaten raw wild boar liver together and had developed severe acute hepatitis E, which was fatal in one of them (36). HEV RNA was detected in the serum obtained from one of them; however, no leftover meat was available. More direct evidence was provided by a report of a series of seven patients with HEV infection in Japan who were among the people from two closely related families who had eaten uncooked deer meat 6 to 7 weeks before the onset of illness (52). A leftover portion of the deer meat, kept frozen, tested positive for HEV RNA, which had nucleotide sequences identical to those of HEV RNA isolated from the patients. Subsequently, more hepatitis E cases related to ingestion of undercooked or uncooked meat in Japan have been reported. In another Japanese study, anti-HEV antibodies were found more often among healthy persons who reported having ever eaten raw deer meat than in those who had never done so.

HEV RNA was shown to be present in 7 of 363 raw pig liver packages available for sale in grocery stores in Hokkaido, Japan (60); on partial sequence analysis, the isolates belonged to HEV genotypes 3 or 4, and some of these had 98.5 to 100% sequence identity to HEV isolates from patients with infections previously reported from the same city. In a recent study, similar data were reported from the United States. In this study, of the 127 pig livers tested, 14 were positive for HEV RNA; all these isolates belonged to HEV genotype 3 (17). Inoculation of homogenates of these pig livers into naïve pigs led to evidence of infection in two of three animals, indicating the presence of infectious virus.

In another study from Japan, nucleic acid sequences from genotype 3 HEV were detected from 2 of 32 packages of a bivalve (*Corbicula japonica*) collected from Japanese rivers (33); however, no transmission studies to determine infectivity were done. Transmission of acute hepatitis E after ingestion of a Chinese herbal medicine by a Japanese man has also been reported (25). All these data confirm that food-borne transmission of HEV infection, particularly from meat of animals that are infected with HEV, can occur.

Differences between HEV Disease in Areas with and without Endemic Infection

There are some differences between patients with indigenous cases of hepatitis E reported from areas where HEV disease is not endemic and cases reported from areas where it is endemic. The former have in general been older and more likely to have other, coexisting diseases and have often had poor outcomes. These cases have been related to infection with genotype 3 HEV (56); cases among pregnant women have not been reported. In contrast, patients in HEV-endemic regions are often young, do not have a coexisting disease, and generally do well, except for pregnant women, who are prone to severe disease.

These findings suggest that epidemiology of HEV infection may be different in the disease-endemic and nonendemic areas and that the genotype of the infecting HEV strain may influence the nature of the clinical disease. In HEV-endemic areas, contamination of drinking-water supplies is common due to poor general sanitation and poor attention to personal hygiene. This may result in easy spread of genotype 1 or 2 HEV from an environmental, human, or animal reservoir, resulting in large outbreaks and sporadic cases, as well as frequent subclinical HEV infection, explaining the occurrence of anti-HEV antibodies in the absence of a history of hepatitis-like illness. On the other hand, in areas where infection is not endemic, genotypes 1 and 2 HEV have possibly been eliminated due to better sanitation, good water quality, and sufficient attention to personal hygiene. However, a large-animal reservoir of genotype 3 HEV still exists, and the occasional animal-to-human transmission from this source may explain the small number of cases seen in these regions. Host species tropism of the genotype 3 virus may account for its failure to cause disease among healthier young adults and its predilection to cause disease among elderly people with other diseases and/or weaker immune response, which in turn may explain the poorer outcome. However, further studies are needed to clarify these issues.

Prevention of Food-Borne Transmission

Only limited data are available on methods for detection of HEV in food items. The only available data are from regions where HEV infection is not endemic. Thus, as referred to above, HEV RNA has been detected, using nucleic acid amplification techniques, in leftover frozen meat that was implicated in causing a small outbreak and in pig liver meat bought from supermarkets in Japan. Based on the sensitivity of the various methods available for the detection of HEV or its components, molecular tests may be theoretically expected to be the most appropriate for this purpose. However, further studies of the amount of viral inoculum required for the development of disease and the sensitivity of the HEV RNA amplification assays may be needed before one can presume that the available assays are sensitive enough to assess the infectivity of food substances.

Adequate heating and proper handling of foods may be expected to reduce the risk of food-borne transmission of HEV infection, although direct evidence supporting the efficacy of these measures is lacking.

In a recent study, administration of three doses of a recombinant HEV vaccine was shown to have a high protective efficacy against clinical hepatitis E in an HEV-endemic region (48). However, the protective efficacy was somewhat lower after two doses, and no data were provided on the protective efficacy of a single dose. Thus, the role of this vaccine in postexposure prophylaxis remains unclear. Further data are also needed on the duration of protection induced by this vaccine before its exact role in the prevention of HEV infection can be assessed.

OTHER ENTERICALLY TRANSMITTED HEPATITIS VIRUSES

A proportion of cases of acute viral hepatitis around the world appear not to be caused by any of the currently known hepatitis viruses. This suggests the possibility that one or more as yet unidentified hepatotropic viruses might exist.

Attempts at identifying novel viruses responsible for parenterally transmitted hepatitis led to the identification of genetic material of several putative viruses, such as hepatitis G virus/GB virus, TT virus, etc. However, subsequent studies showed that although these viruses were transmitted parenterally, they were not responsible for liver disease.

Similar attempts at identification of novel enterically transmitted hepatitis viruses have been much more limited. In a retrospective seroepidemiological study of 17 epidemics of waterborne hepatitis in India, one epidemic in the Andaman Islands appeared to have been caused by a previously unrecognized hepatitis agent (2). In the other 16 epidemics, a significant minority of cases lacked serological markers of HEV infection. In addition, in the HEV-endemic regions, known viral agents can account for a fairly large proportion of cases of acute sporadic hepatitis. There is thus a need for more work in this field.

CONCLUSIONS

HAV and HEV are common causes of enterically transmitted viral hepatitis. Both these viruses have RNA genomes, are excreted in feces of the infected persons, and are transmitted by the fecal-oral route. Transmission of infection through contamination of food has been studied well for HAV. In contrast, outbreaks of HEV infection have been more often related to water contamination. Molecular techniques are being used to determine the potential of various food items to cause hepatitis A. This seems to be possible for hepatitis E too, although data on this infection are relatively limited.

REFERENCES

1. **Aggarwal, R., and S. R. Naik.** 1994. Hepatitis E: intrafamilial transmission versus waterborne spread. *J. Hepatol.* **21:**718–723.

2. **Arankalle, V. A., M. S. Chadha, S. A. Tsarev, S. U. Emerson, A. R. Risbud, K. Banerjee, and R. H. Purcell.** 1994. Seroepidemiology of water-borne hepatitis in India and evidence for a third enterically-transmitted hepatitis agent. *Proc. Natl. Acad. Sci. USA* **91:**3428–3432.

3. **Arankalle, V. A., L. P. Chobe, M. V. Joshi, M. S. Chadha, B. Kundu, and A. M. Walimbe.** 2002. Human and swine hepatitis E viruses from Western India belong to different genotypes. *J. Hepatol.* **36:**417–425.

4. **Beller, M.** 1992. Hepatitis A outbreak in Anchorage, Alaska, traced to ice slush beverages. *West. J. Med.* **156:**624–627.

5. **Brack, K., W. Frings, A. Dotzauer, and A. Vallbracht.** 1998. A cytopathogenic, apoptosis-inducing variant of hepatitis A virus. *J. Virol.* **72:**3370–3376.

6. **Bradley, D., A. Andjaparidze, E. H. Cook, Jr., K. McCaustland, M. Balayan, H. Stetler, O. Velazquez, B. Robertson, C. Humphrey, M. Kane, and I. Weisfuse.** 1988. Aetiological agent of enterically transmitted non-A, non-B hepatitis. *J. Gen. Virol.* **69:**731–738.

7. **Calci, K. R., G. K. Meade, R. C. Tezloff, and D. H. Kingsley.** 2005. High-pressure inactivation of hepatitis A virus within oysters. *Appl. Environ. Microbiol.* **71:**339–343.

7a. **Chen, H.-Y., Y. Lu, T. Howard, D. Anderson, P. Y. Fong, W.-P. Hu, C. P. Chia, and M. Guan.** 2005. Enzyme-linked immunosorbent assay for rapid detection of immunoglobulin M antibodies to hepatitis E virus in human sera. *Clin. Diagn. Lab. Immunol.* **12:**593–598.

8. Cohen, J. I., J. R. Ticehurst, R. H. Purcell, A. Buckler-White, and B. M. Baroudy. 1987. Complete nucleotide sequence of wild-type hepatitis A virus: comparison with different strains of hepatitis A virus and other picornaviruses. *J. Virol.* **61**:50–59.

9. Conaty, S., P. Bird, G. Bell, E. Kraa, G. Grohmann, and J. M. McAnulty. 2000. Hepatitis A in New South Wales, Australia, from consumption of oysters: the first reported outbreak. *Epidemiol. Infect.* **124**:121–130.

10. Dawson, G. J., K. H. Chau, C. M. Cabal, P. O. Yarbough, G. R. Reyes, and I. K. Mushahwar. 1992. Solid-phase enzyme-linked immunosorbent assay for hepatitis E virus IgG and IgM antibodies utilizing recombinant antigens and synthetic peptides. *J. Virol. Methods* **38**:175–186.

11. De Medici, D., M. Ciccozzi, A. Fiore, S. Di Pasquale, A. Parlato, P. Ricci-Bitti, and L. Croci. 2001. Closed-circuit system for the depuration of mussels experimentally contaminated with hepatitis A virus. *J. Food Prot.* **64**:877–880.

12. Desenclos, J. C., K. C. Klontz, M. H. Wilder, O. V. Nainan, H. S. Margolis, and R. A. Gunn. 1991. A multistate outbreak of hepatitis A caused by the consumption of raw oysters. *Am. J. Public Health* **81**:1268–1272.

13. Emerson, S. U., V. A. Arankalle, and R. H. Purcell. 2005. Thermal stability of hepatitis E virus. *J. Infect. Dis.* **192**:930–933.

14. Emerson, S. U., P. Clemente-Casares, N. Moiduddin, V. A. Arankalle, U. Torian, and R. H. Purcell. 2006. Putative neutralization epitopes and broad cross-genotype neutralization of hepatitis E virus confirmed by a quantitative cell-culture assay. *J. Gen. Virol.* **87**:697–704.

15. Emerson, S. U., H. Nguyen, J. Graff, D. A. Stephany, A. Brockington, and R. H. Purcell. 2004. In vitro replication of hepatitis E virus (HEV) genomes and of an HEV replicon expressing green fluorescent protein. *J. Virol.* **78**:4838–4846.

16. Enriquez, R., G. G. Frosner, V. Hochstein-Mintzel, S. Riedemann, and G. Reinhardt. 1992. Accumulation and persistence of hepatitis A virus in mussels. *J. Med. Virol.* **37**:174–179.

17. Feagins, A. R., T. Opriessnig, D. K. Guenette, P. G. Halbur, and X. J. Meng. 2007. Detection and characterization of infectious hepatitis E virus from commercial pig livers sold in local grocery stores in the USA. *J. Gen. Virol.* **88**:912–917.

18. Hadler, S. C., H. M. Webster, J. J. Erben, J. E. Swanson, and J. E. Maynard. 1980. Hepatitis A in day-care centers. A community-wide assessment. *N. Engl. J. Med.* **302**:1222–1227.

19. Halbur, P. G., C. Kasorndorkbua, C. Gilbert, D. Guenette, M. B. Potters, R. H. Purcell, S. U. Emerson, T. E. Toth, and X. J. Meng. 2001. Comparative pathogenesis of infection of pigs with hepatitis E viruses recovered from a pig and a human. *J. Clin. Microbiol.* **39**:918–923.

20. Halliday, M. L., L. Y. Kang, T. K. Zhou, M. D. Hu, Q. C. Pan, T. Y. Fu, Y. S. Huang, and S. L. Hu. 1991. An epidemic of hepatitis A attributable to the ingestion of raw clams in Shanghai, China. *J. Infect. Dis.* **164**:852–859.

21. Huang, F. F., Z. F. Sun, S. U. Emerson, R. H. Purcell, H. L. Shivaprasad, F. W. Pierson, T. E. Toth, and X. J. Meng. 2004. Determination and analysis of the complete genomic sequence of avian hepatitis E virus (avian HEV) and attempts to infect rhesus monkeys with avian HEV. *J. Gen. Virol.* **85**:1609–1618.

22. Hutin, Y. J., V. Pool, E. H. Cramer, O. V. Nainan, J. Weth, I. T. Williams, S. T. Goldstein, K. F. Gensheimer, B. P. Bell, C. N. Shapiro, M. J. Alter, and H. S. Margolis for the National Hepatitis A Investigation Team. 1999. A multistate, foodborne outbreak of hepatitis A. *N. Engl. J. Med.* **340:**595–602.

23. Hutin, Y. J., K. M. Sabin, L. C. Hutwagner, L. Schaben, G. M. Shipp, D. M. Lord, J. S. Conner, M. P. Quinlisk, C. N. Shapiro, and B. P. Bell. 2000. Multiple modes of hepatitis A virus transmission among methamphetamine users. *Am. J. Epidemiol.* **152:**186–192.

24. Innis, B. L., R. Snitbhan, P. Kunasol, T. Laorakpongse, W. Poopatanakool, C. A. Kozik, S. Suntayakorn, T. Suknuntapong, A. Safary, D. B. Tang, et al. 1994. Protection against hepatitis A by an inactivated vaccine. *JAMA* **271:**1328–1334.

25. Ishikawa, K., K. Matsui, T. Madarame, S. Sato, K. Oikawa, and T. Uchida. 1995. Hepatitis E probably contracted via a Chinese herbal medicine, demonstrated by nucleotide sequencing. *J. Gastroenterol. Hepatol.* **30:**534–538.

26. Khuroo, M. S. 1980. Study of an epidemic of non-A, non-B hepatitis. Possibility of another human hepatitis virus distinct from post-transfusion non-A, non-B type. *Am. J. Med.* **68:**818–824.

27. Khuroo, M. S., S. Kamili, M. Y. Dar, R. Moecklii, and S. Jameel. 1993. Hepatitis E and long-term antibody status. *Lancet* **341:**1355.

28. Khuroo, M. S., S. Kamili, and S. Jameel. 1995. Vertical transmission of hepatitis E virus. *Lancet* **345:**1025–1026.

29. Khuroo, M. S., S. Kamili, and G. N. Yattoo. 2004. Hepatitis E virus infection may be transmitted through blood transfusions in an endemic area. *J. Gastroenterol. Hepatol.* **19:**778–784.

30. Kumar, A., R. Aggarwal, S. R. Naik, V. Saraswat, U. C. Ghoshal, and S. Naik. 2004. Hepatitis E virus is responsible for decompensation of chronic liver disease in an endemic region. *Indian J. Gastroenterol.* **23:**59–62.

31. Lednar, W. M., S. M. Lemon, J. W. Kirkpatrick, R. R. Redfield, M. L. Fields, and P. W. Kelley. 1985. Frequency of illness associated with epidemic hepatitis A virus infections in adults. *Am. J. Epidemiol.* **122:**226–233.

32. Lemon, S. M. 1993. Immunologic approaches to assessing the response to inactivated hepatitis A vaccine. *J. Hepatol.* **18**(Suppl. 2):S15–S19.

33. Li, T. C., T. Miyamura, and N. Takeda. 2007. Detection of hepatitis E virus RNA from the bivalve Yamato-Shijimi (*Corbicula japonica*) in Japan. *Am. J. Trop. Med. Hyg.* **76:**170–172.

34. Lowry, P. W., R. Levine, D. F. Stroup, R. A. Gunn, M. H. Wilder, and C. Konigsberg, Jr. 1989. Hepatitis A outbreak on a floating restaurant in Florida, 1986. *Am. J. Epidemiol.* **129:**155–164.

35. Martin, A., and S. M. Lemon. 2006. Hepatitis A virus: from discovery to vaccines. *Hepatology* **43:**S164–S172.

36. Matsuda, H., K. Okada, K. Takahashi, and S. Mishiro. 2003. Severe hepatitis E virus infection after ingestion of uncooked liver from a wild boar. *J. Infect. Dis.* **188:**944.

37. Mbithi, J. N., V. S. Springthorpe, J. R. Boulet, and S. A. Sattar. 1992. Survival of hepatitis A virus on human hands and its transfer on contact with animate and inanimate surfaces. *J. Clin. Microbiol.* **30:**757–763.

38. McCaustland, K. A., W. W. Bond, D. W. Bradley, J. W. Ebert, and J. E. Maynard. 1982. Survival of hepatitis A virus in feces after drying and storage for 1 month. *J. Clin. Microbiol.* **16:**957–958.

39. Meng, X.-J., P. G. Halbur, M. S. Shapiro, S. Govindarajan, J. D. Bruna, I. K. Mushahwar, R. H. Purcell, and S. U. Emerson. 1998. Genetic and experimental evidence for cross-species infection by swine hepatitis E virus. *Am. Soc. Microbiol.* **72:**9714–9721.

40. Meng, X. J., R. H. Purcell, P. G. Halbur, J. R. Lehman, D. M. Webb, T. S. Tsareva, J. S. Haynes, B. J. Thacker, and S. U. Emerson. 1997. A novel virus in swine is closely related to the human hepatitis E virus. *Proc. Natl. Acad. Sci. USA* **94:**9860–9865.

41. Naik, S. R., R. Aggarwal, P. N. Salunke, and N. N. Mehrotra. 1992. A large waterborne viral hepatitis E epidemic in Kanpur, India. *Bull. W. H. O.* **70:**597–604.

42. Nainan, O. V., G. Xia, G. Vaughan, and H. S. Margolis. 2006. Diagnosis of hepatitis A virus infection: a molecular approach. *Clin. Microbiol. Rev.* **19:**63–79.

43. Nicand, E., G. L. Armstrong, V. Enouf, J. P. Guthmann, J. P. Guerin, M. Caron, J. Y. Nizou, and R. Andraghetti. 2005. Genetic heterogeneity of hepatitis E virus in Darfur, Sudan, and neighboring Chad. *J. Med. Virol.* **77:**519–521.

44. Nuanualsuwan, S., T. W. Mariam, S. Himathongkham, and D. O. Cliver. 2002. Ultraviolet inactivation of feline calicivirus, human enteric viruses and coliphages. *Photochem. Photobiol.* **76:**406–410.

45. Robertson, B. H., R. W. Jansen, B. Khanna, A. Totsuka, O. V. Nainan, G. Siegl, A. Widell, H. S. Margolis, S. Isomura, K. Ito, T. Ishizu, Y. Moritsugu, and S. M. Lemon. 1992. Genetic relatedness of hepatitis A virus strains recovered from different geographic regions. *J. Gen. Virol.* **73:**1365–1377.

46. Schlauder, G. G., B. Frider, S. Sookoian, G. C. Castano, and I. K. Mushahwar. 2000. Identification of 2 novel isolates of hepatitis E virus in Argentina. *J. Infect. Dis.* **182:**294–297.

47. Shan, X. C., P. Wolffs, and M. W. Griffiths. 2005. Rapid and quantitative detection of hepatitis A virus from green onion and strawberry rinses by use of real-time reverse transcription-PCR. *Appl. Environ. Microbiol.* **71:**5624–5626.

48. Shrestha, M. P., R. M. Scott, D. M. Joshi, M. P. Mammen, Jr., G. B. Thapa, N. Thapa, K. S. Myint, M. Fourneau, R. A. Kuschner, S. K. Shrestha, M. P. David, J. Seriwatana, D. W. Vaughn, A. Safary, T. P. Endy, and B. L. Innis. 2007. Safety and efficacy of a recombinant hepatitis E vaccine. *N. Engl. J. Med.* **356:**895–903.

49. Somani, S. K., R. Aggarwal, S. R. Naik, S. Srivastava, and S. Naik. 2003. A serological study of intrafamilial spread from patients with sporadic hepatitis E virus infection. *J. Viral Hepat.* **10:**446–449.

50. Staes, C. J., T. L. Schlenker, I. Risk, K. G. Cannon, H. Harris, A. T. Pavia, C. N. Shapiro, and B. P. Bell. 2000. Sources of infection among persons with acute hepatitis A and no identified risk factors during a sustained community-wide outbreak. *Pediatrics* **106:**E54.

51. Tanaka, T., M. Takahashi, E. Kusano, and H. Okamoto. 2007. Development and evaluation of an efficient cell-culture system for hepatitis E virus. *J. Gen. Virol.* **88:**903–911.

52. Tei, S., N. Kitajima, K. Takahashi, and S. Mishiro. 2003. Zoonotic transmission of hepatitis E virus from deer to human beings. *Lancet* **362:**371–373.

53. Tsarev, S. A., S. U. Emerson, G. R. Reyes, T. S. Tsareva, L. J. Legters, I. A. Malik, M. Iqbal, and R. H. Purcell. 1992. Characterization of a prototype strain of hepatitis E virus. *Proc. Natl. Acad. Sci. USA* **89:**559–563.

54. Velazquez, O., H. C. Stetler, C. Avila, G. Ornelas, C. Alvarez, S. C. Hadler, D. W. Bradley, and J. Sepulveda. 1990. Epidemic transmission of enterically transmitted non-A, non-B hepatitis in Mexico, 1986–1987. *JAMA* **263**:3281–3285.

55. Viswanathan, R. 1957. Infectious hepatitis in Delhi (1955–56). A critical study: epidemiology. *Indian J. Med. Res.* **45**:1–29.

56. Waar, K., M. M. Herremans, H. Vennema, M. P. Koopmans, and C. A. Benne. 2005. Hepatitis E is a cause of unexplained hepatitis in The Netherlands. *J. Clin. Virol.* **33**:145–149.

57. Werzberger, A., B. Mensch, B. Kuter, L. Brown, J. Lewis, R. Sitrin, W. Miller, D. Shouval, B. Wiens, G. Calandra, J. Ryan, P. Provost, and D. Nalin. 1992. A controlled trial of a formalin-inactivated hepatitis A vaccine in healthy children. *N. Engl. J. Med.* **13**:453–457.

58. Wheeler, C., T. M. Vogt, G. L. Armstrong, G. Vaughan, A. Weltman, O. V. Nainan, V. Dato, G. Xia, K. Waller, J. Amon, T. M. Lee, A. Highbaugh-Battle, C. Hembree, S. Evenson, M. A. Ruta, I. T. Williams, A. E. Fiore, and B. P. Bell. 2005. An outbreak of hepatitis A associated with green onions. *N. Engl. J. Med.* **353**:890–897.

59. Willner, I. R., M. D. Uhl, S. C. Howard, E. Q. Williams, C. A. Riely, and B. Waters. 1998. Serious hepatitis A: an analysis of patients hospitalized during an urban epidemic in the United States. *Ann. Intern. Med.* **128**:111–114.

60. Yazaki, Y., H. Mizuo, M. Takahashi, T. Nishizawa, N. Sasaki, Y. Gotanda, and H. Okamoto. 2003. Sporadic acute or fulminant hepatitis E in Hokkaido, Japan, may be food-borne, as suggested by the presence of hepatitis E virus in pig liver as food. *J. Gen. Virol.* **84**:2351–2357.

61. Yotsuyanagi, H., K. Koike, K. Yasuda, K. Moriya, Y. Shintani, H. Fujie, K. Kurokawa, and S. Iino. 1996. Prolonged fecal excretion of hepatitis A virus in adult patients with hepatitis A as determined by polymerase chain reaction. *Hepatology* **24**:10–13.

62. Zhuang, H., X.-Y. Cao, C.-B. Liu, and G.-M. Wang. 1991. Enterically transmitted non-A, non-B hepatitis in China, p. 277–285. *In* T. Shikata, R. H. Purcell, and T. Uchida (ed.), *Viral Hepatitis C, D and E.* Excerpta Medica, Amsterdam, The Netherlands.

Food-Borne Viruses: Progress and Challenges
Edited by Marion P. G. Koopmans, Dean O. Cliver, and Albert Bosch
© 2008 ASM Press, Washington, DC

The Challenge of Estimating the Burden of an Underreported Disease

4

Sarah J. O'Brien

Diarrhea is a comedy illness. It is unpleasant for the victim but is not generally regarded as serious or life threatening. Its perception as a trivial illness perhaps poses the greatest challenge in developing robust estimates of disease burden. This chapter focuses on the epidemiologic and surveillance challenges posed by food-borne viral disease. The exemplar used is norovirus (NoV), although most of the principles apply equally to other viruses transmitted through food.

CHALLENGES IN THE SURVEILLANCE OF FOOD-BORNE VIRAL DISEASE

Describing the burden of any disease in the population usually depends on some form of surveillance system, but the cases reported to such systems are not necessarily representative of the entire population. Wise interpretation of surveillance data involves understanding the points at which data are lost and the information bias that results from these losses.

The Virus and the Patient

One of the main challenges for undertaking surveillance of food-borne viruses, especially NoV, is posed by the viruses themselves. Most ongoing surveillance schemes depend on patients consulting with or presenting to a primary-care physician. Without this step, the illness is unlikely to be recorded in any official statistics. Many food-borne viral infections are relatively short-lived, but patients who consult their primary-care physician tend to perceive their

SARAH J. O'BRIEN, School of Translational Medicine, University of Manchester, Clinical Sciences Building, Hope Hospital, Stott Lane, Salford M6 8HD, United Kingdom.

symptoms as more severe and/or, are more likely to have returned recently from a trip overseas, to present with bloody diarrhea (not a feature of NoV infection), or to have prolonged symptoms (19, 76, 84). In population-based studies of acute gastroenteritis, only 5 to 20% of those who develop symptoms of acute gastroenteritis (AGE) consult a primary-care physician or other health care professional (19, 75, 76, 85). These findings are also echoed in a survey of self-care of common disorders conducted in Spain (60). Self-management was common among the population studied for a variety of frequently reported symptoms including headache, the common cold, backache, and diarrhea.

The wide variation in consultation behavior is illustrated by two population-based studies. In England 1 in 23 patients consulted their primary-care physician following an episode of AGE, whereas in The Netherlands 1 in 125 patients did so (22, 93). Thus, the greatest potential loss of information about cases and hence about disease burden occurs before official surveillance systems are ever even triggered.

The Physician and the Health Care System

If a patient does consult a primary-care physician, the potential for data capture depends on the clinical decision-making of the physician. Should the physician decide not to ask the patient for a stool sample, valuable diagnostic information is lost. In a study in British Columbia, physicians were less likely to request stool samples from patients experiencing vomiting, a major symptom in NoV infection (54), than to request samples for illnesses of long duration or greater severity or when the patient had recently returned from abroad (74, 90). From the physician's point of view, there is probably very little to be gained by ordering laboratory tests since they are unlikely to alter patient management. However, from a public health point of view, the diagnosis can be extremely important since management differs depending on the causative pathogen.

The patient's response to a request to submit a stool sample is another point at which valuable information is lost. Only one-fifth to one-third of patients asked by their physician to provide a stool sample actually do so (75, 76).

Certain countries have introduced nurse-led telephone triage systems to provide the first line of advice for patients and to channel their route through the health care system. In a study of calls to a nurse-led telephone triage system in the United Kingdom, around 10% of over 152,000 calls to the system in a 6-month period concerned gastrointestinal illness (i.e., diarrhea, vomiting, or "food poisoning"), amounting to 8 calls per 100,000 population per week (14). When the outcome of all calls was examined, a significantly higher proportion of patients calling about gastrointestinal symptoms were advised about management at home (41.5%; 95% confidence interval, 40.5 to 42.5) than were those calling about other symptoms (23.4%; 95% confidence interval, 23.1 to 23.7). The likelihood of being referred urgently to a primary-care physician depended on the age of the patient, with children being more likely to be referred. To understand better the selection bias that occurs when people self-referred to the telephone triage system, Cooper et al. looked at the effect of age, sex, and deprivation on call rates to the system (13). Call rates were higher in areas of deprivation close to or above the national average, showing that demand is higher in areas of greater deprivation. Call rates were higher when enquiring about symptoms in children younger than 5 years, but, paradoxically, calls about children were reduced in areas of extreme deprivation.

The Laboratory

Assuming that a sample from a symptomatic patient is requested and obtained, the next stage at which diagnostic information is lost is in the laboratory. The diagnostic testing algorithms and tests employed will determine the likelihood of a viral diagnosis being obtained. Not all primary diagnostic laboratories routinely examine stool samples for the presence of NoV, and some do so only if an outbreak is suspected. Those that do examine samples use a variety of methods, including electron microscopy (EM) (high specificity but low sensitivity), enzyme-linked immunosorbent assays (often used as a first-line screen in laboratories that test for NoV since they detect the commonly circulating strains), and reverse transcription-PCR assays (RT-PCR) (high sensitivity).

However, a negative laboratory finding does not necessarily mean that NoV is absent. Nowhere is this better illustrated than in a study examining archived DNA and cDNA from stool samples from a large population-based case-control study of infectious intestinal disease (IID) in England. In the original study, stools were examined for NoV by EM, and 6.5% of cases (169 of 2,893) and 0.3% of controls (6 of 2,210) were found to be infected with NoV (86). Recently, Amar et al. applied PCR assays to DNA and cDNA obtained from 4,627 fecal samples during the original study (3). They reanalyzed the samples by the EM-based methods originally employed in the IID study as well as by the PCR assays. When EM alone was used (as in the original study), NoV was detected in 6% of samples from cases; however, this increased to 36% when RT-PCR was used. In controls the proportion testing positive increased from 2.7% with EM alone to 16% when RT-PCR was employed as well. In this study, the increased detection of NoV was statistically significant in all age groups. The use of PCR significantly increased the number of individuals from whom more than one pathogen was detected from 272 to 993 (73% increase) in cases and from 32 to 280 (89% increase) in controls, compared with the results of conventional testing. Among cases, single and multiple agents were detected in 36 and 41%, respectively. NoV was the agent most often detected on its own, especially in individuals aged 5 to 19, 30 to 49, and over 70 years. However, the proportion of multiple-agent over single-agent detections involving NoV was the highest in those younger than 5 years. This study illustrates dramatically that what you find in stool samples depends not only on what you look for but also how hard you look for it.

However, this study also poses several challenges. The first is that while the increase in sensitivity afforded by PCR is undeniable, its specificity is less certain. Quantitative assays are needed before we can say with some degree of certainty that the presence of genetic material in stool samples means that the detected agent caused the disease. The second is the interpretation of multiple-agent findings. Which of the multiple pathogens present caused the patients' illnesses? It is possible that quantitative assays might be of benefit here as well. The third challenge is that the high proportion of controls in whom NoV was found might mean that controls in the original study were misclassified. Should they have been cases after all? Shedding of virus has been demonstrated in 26% of patients up to 3 weeks after illness onset (70), lending further weight to the potential for misclassification of cases and controls in epidemiologic studies.

The Public Health System

There are at least three ways in which the public health system can militate against a good understanding of the burden of food-borne viral gastroenteritis. First, there must be a commitment to, and process for, collating data on cases or outbreaks of food-borne viral gastroenteritis. This is not a requirement in all countries, so that surveillance of food-borne viral gastroenteritis is at best patchy. Second, the approach to investigation of sporadic cases of food-borne disease is not uniform. For example, in a survey of environmental health departments (sanitarians) in England, 64% of respondents always investigated a case of viral gastroenteritis reported to them (compared with 96% for *Salmonella* enteritis) (71). In general, cases of bacterial gastroenteritis were much more likely to be investigated (with the exception of *Campylobacter* infection); however, since 13% of respondents waited until they had received laboratory confirmation of a sporadic case of food poisoning, the proportion of cases of sporadic viral gastroenteritis investigated is likely to be overestimated in this survey. Finally, for an infection that is, perhaps, classically regarded as predominantly outbreak related, the approach to the investigation of outbreaks is also variable. In a review of 1,426 food-borne general outbreaks of IID in England and Wales, the evidence to support a food-borne route of transmission was supplied for two-thirds of the outbreaks investigated (61). Overall, a combination of microbiological and analytical evidence was reported in 4% of outbreaks (60 of 1,426), microbiological evidence alone in 10% (149 of 1,426), analytical evidence alone in 23% (322 of 1,426), microbiological evidence in combination with descriptive epidemiology in 3% (46 of 1,426), and descriptive epidemiology alone in 26% (365 of 1,426). However, for NoV outbreaks thought to be food borne, which made up 6% (83 of 1,426) of the total, a combination of microbiological and analytical evidence was reported in 1% of outbreaks (1 of 83), microbiological evidence alone in 1% (1 of 83), analytical evidence alone in 36% (30 of 83), microbiological evidence in combination with descriptive epidemiology in 2% (2 of 83), and descriptive epidemiology alone in 23% (19 of 83); in 36% of the outbreaks of NoV infection, the evidence supporting food-borne transmission was not reported. It is striking, but not surprising, how the lack of good tools for detecting viruses in food has affected outbreak investigations. This has placed greater emphasis on analytical epidemiologic evidence to implicate food vehicles in NoV outbreaks. However, the chance to identify a contaminated food vehicle was lost in a substantial proportion of outbreaks.

The Surveillance Pyramid

The attrition of data at various points along the surveillance chain from patient to official statistics is often described as a pyramid (Fig. 1). Disease in the community forms the base of the pyramid, while cases that reach official statistics form the apex. Some researchers have attempted to equate disease in the population to what appears in official statistics for acute gastroenteritis, but few have attempted this specifically for NoV.

In the IID study in England in the mid-1990s, the investigators determined that for every case of IID reported to national surveillance, 136 occurred in the community (93), while in British Columbia, for every case of infectious gastrointestinal illness reported to the province, there was a mean of 347 cases in the community (54). When Wheeler et al. (93) looked at the corresponding ratios for NoV, they found that for every case of NoV in the national surveillance data set, there were 1,562 in the community! The figure shows the ratios of IID or infectious gastrointestinal illness at different levels of the pyramid in two international studies. However, it should be noted that the IID study measured rates directly at each step of the pyramid while the study from British Columbia is based on simulation and that the two studies used different case definitions.

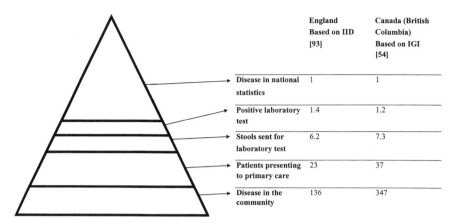

Figure 1 Relationship between disease in the community, presenting to primary care, and reported to national surveillance (ratios).

SURVEILLANCE SYSTEMS—SHEDDING LIGHT ON DISEASE BURDEN?

Despite all the shortcomings described in the previous section, several methods of surveillance are employed for acute gastroenteritis or NoV infection. How far these shed light on the burden of disease is, however, debatable.

Syndromic Surveillance

Syndromic surveillance systems do not necessarily capture information on disease etiology. However, since most patients with mild gastroenteritis do not present to primary-care physicians, and since the majority of cases in community-based studies in developed countries are caused by NoV, it might be argued that syndromic systems contain some useful proxy data on the burden of NoV in the population. There has been rapid growth in the use of such systems following the infamous event of 11 September 2001, when the need to detect unusual events rapidly heightened.

Several potential sources of data exist. In France, syndromic surveillance is based on a voluntary sentinel network of primary-care physicians distributed across the country (25). The physicians report 12 conditions, including acute diarrheas, on a weekly basis by updating a Web-accessible database. An added benefit of this network is the potential to add microbiological investigations as part of targeted studies.

Pharmacy-based syndromic surveillance has been employed in Canada to monitor over-the-counter purchases of medication including antiemetics, antidiarrheals, and rehydration therapies by patients with AGE (31). A total of 31% of 351 self-reported patients with AGE purchased at least one of these medications during their illness. Patients with symptoms of both diarrhea and vomiting were 3.6 times more likely to purchase an over-the-counter medication than were patients with either symptom alone.

In Arizona, a poison control center database was scrutinized to identify callers with gastrointestinal symptoms attributable to suspected food-borne illness (17). The calls were also compared with those to the county health department. There was virtually no overlap between the two data sets, so that capture-recapture analysis would be a valid method for determining the true burden of gastrointestinal symptoms in that population.

Syndromic surveillance of gastrointestinal symptoms has also been used to determine if outbreaks can be identified (5, 15, 80). In two studies, one based on calls to a national health help line in the United Kingdom and one based on presentations to emergency rooms in New York City, the investigators concluded that substantial changes in disease presentation were detectable but more localized events were unlikely to be revealed (5, 15). However, when citywide signals were detected through emergency room surveillance in New York City, targeted public health investigations revealed seasonal NoV to be the cause of the symptoms (80).

A major issue in the success of syndromic surveillance systems is the need to maintain motivation among the participants and ensure compliance with reporting. Standardizing the clinical terms used to report data is also essential (44), as is ensuring that data coding is consistent (80). Finally, syndromic systems tend to be highly sensitive but lack specificity. The sensitivity can be problematic, generating many false-positive signals.

Laboratory-Based Surveillance

Considered here for the sake of completeness, although it is often considered the mainstay for surveillance of bacterial causes of gastroenteritis, routine laboratory-based surveillance is probably the least useful system for assessing the disease burden due to food-borne viruses. There is massive underreporting of infections by food-borne viruses, since patients rarely present to primary-care physicians, and diagnostic testing in many laboratories is outbreak based. This means that this type of surveillance is severely limited as a credible source of information on sporadic infection. However, in targeted studies where stool samples submitted to diagnostic and/or public health laboratories have been examined for NoV, prevalence estimates vary considerably —from 3.3% in South Africa to 35% in Japan (Table 1). Direct comparisons between the studies are fraught with difficulties. As can be seen from Table 1, the sampling frames and detection methods used in the various studies are different and some of the studies also include outbreak cases. Many of the studies are based on infections in children and often do not focus on diarrhea

Table 1 Recent prevalence estimates for "sporadic" NoV infection

Yr(s) of study	Country	Prevalence estimate %	Method(s): Sampling frame	Reference(s)
2004–2005	Russia	9.7	RT-PCR: stools from adults admitted to infectious–diseases department	72
2003–2004	Russia	4	RT-PCR: stools from children with AGE	67
2003	Italy	18.6	EIA[a] and RT-PCR: stools from children admitted to hospital with AGE	12
2002	Italy	10.4	RT-PCR: stools from children with sporadic pediatric gastroenteritis admitted to hospital	57
2001–2003	Switzerland	17.9	RT-PCR: negative stools from individuals with community-acquired diarrhea	27
2001–2002	Hong Kong	9.2 (surveillance specimens); 16.7 (clinical specimens); 82.8 (outbreak specimens)	RT-PCR: stools from three different sources (community surveillance, hospital patients, with AGE, individuals in outbreaks)	50
2000–2003	Japan	35	RT-PCR: stools from children with sporadic gastroenteritis diagnosed in sentinel pediatric clinics	82
2000–2001	Spain	12.9	RT-PCR: stools from children with sporadic cases of pediatric AGE (< 5 years of age)	10
1999–2004	Japan	31.8	RT-PCR: stools from hospital patients and outbreak investigations	64
1999–2000	Vietnam	5.4	RT-PCR: rotavirus-negative stools from hospitalized children	37
Not stated	China (Guangzhou Province)	4 (cases of nonbacterial gastroenteritis); 30% (elementary school outbreak)	RT-PCR: stools from children	16
1998–2002	Australia	9.2 (prevalence of calicivirus)	RT-PCR: stools from children with pediatric gastroenteritis admitted to hospital	48
1999–2000	England	10.3	EIA, EM, and RT-PCR: stools from children with sporadic pediatric gastroenteritis referred to a public health laboratory	30
1997–1999	United States	7.1	RT-PCR: stool samples from children <5 years admitted to three hospitals	96
1997–1999	Indonesia	20.6	RT-PCR: stools negative on culture and parasitology from patients admitted to or seen at one hospital and three health centers	81

1997–1999	Australia	10.7	RT-PCR: stools from a prospective community-based study of patients presenting with "highly credible gastroenteritis"	78
1996–1998	Ireland	8	RT-PCR: stools from patients with sporadic gastroenteritis referred to a hospital laboratory	26
1998–1999	The Netherlands	5.1 in cases at physician presentation; 11 in community cohort	RT-PCR: stools from individuals presenting to primary-care physicians and a prospective community-based cohort study	20, 21
1993–1996	England	6.5 at physician presentation; 7 in community cohort	EM: stools from individuals presenting to primary-care physicians and a prospective community-based cohort study	86
1991–1995	South Africa	3.3 (human calicivirus)	EIA, EM, and RT-PCR: stools from individuals with sporadic gastroenteritis referred to selected laboratories	95

[a] EIA, enzyme immunoassay.

in the community; thus, selection bias and detection bias cloud the picture. Sample sizes also vary markedly, so that the precision associated with prevalence estimates will differ.

Outbreak Surveillance

Many countries collect data on food-borne disease outbreaks. Since NoV often presents as an outbreak-related disease, surveillance of outbreaks of NoV should provide good insight into the disease burden. But does it? Like routine laboratory-based surveillance, many outbreak surveillance systems tend to be geared toward bacterial pathogens. Moreover, while some surveillance systems capture data on all outbreaks, no matter what the setting, others do not include outbreaks that occur within families because it can be difficult to differentiate co-primary cases from cases caused by secondary spread of infection in families.

Where data on NoV are collected, the predominant mode of transmission tends to be identified as person to person, and health care settings stand out as being most severely affected in outbreaks (8). In a European survey of countries that conduct broad-based outbreak surveillance, the proportions of viral gastroenteritis outbreaks that were associated with food-borne or waterborne transmission were as follows: Finland (24%), The Netherlands (17%), Slovenia (14%), Spain (7%), and England and Wales (7%) (52). The survey showed that laboratory evidence (detection of the same organism in the vehicle and stool specimens) or analytical epidemiologic evidence (from case-control or cohort studies) that demonstrated the association between the suspected food vehicle and illness was very rare.

In the absence of wide use of NoV diagnostics, clinical and epidemiologic criteria (commonly referred to as Kaplan's criteria) (47) are often applied to outbreaks to determine the likelihood of a viral etiology. In a recent review of 4,050 outbreaks reported to the Centers for Disease Control and Prevention, Turcios et al. examined how well clinical and epidemiologic profiles discriminate between food-borne outbreaks of gastroenteritis due to NoV and those due to bacteria, and they estimated the proportion of reported outbreaks that might be attributable to NoV (87). They concluded that Kaplan's criteria were highly specific (99%) and moderately sensitive (68%) in discriminating confirmed outbreaks due to bacteria from those due to NoV, and they determined that, at a minimum, 28% of all the food-borne outbreaks reported could be attributed to NoV on the basis of those criteria. Despite these useful findings, not all surveillance systems capture sufficient clinical or epidemiologic information to be able to apply these criteria as a matter of routine.

Active Surveillance

A number of developed countries have adopted active-surveillance schemes to monitor food-borne disease. FoodNet in the United States and Oz FoodNet in Australia are two such examples. However, although both have used population-based, retrospective, cross-sectional surveys to determine the incidence of AGE, neither specifically includes NoV as one of the target organisms (4, 46).

Integrated Food Chain Surveillance

While human enteric viruses are now recognized as common causes of food-borne disease and methods for detecting these agents in clinical specimens have improved significantly over the last decade or so, applications to food samples have progressed much more slowly (73). When testing has been performed, it has tended to be limited to shellfish commodities (51). Methods to detect human enteric viruses in foods other than shellfish have been developed over recent years (11, 33), but as yet none is capable of being used in systematic surveys of food commodities.

Interpreting Surveillance Data

On a global scale, interpretation of data from all the surveillance systems mentioned above is limited to a greater or lesser extent by the multiplicity of systems from which they are drawn (63). Some national surveillance systems are centralized, while others are federalized. Notifiable or reportable diseases differ from nation to nation. The general consensus is that underreporting is common, although, with few exceptions, the extent to which this affects the numerator is not quantified. Another feature which varies by country is the proportion of the population covered by national surveillance systems. National data sets contain different amounts of information, and few employ standard case definitions. These factors all make it very difficult to make sensible conclusions about the burden of food-borne viral gastroenteritis or to draw valid intercountry comparisons. And, of course, there are enormous data gaps from developing countries.

POPULATION-BASED APPROACHES FOR DETERMINING THE DISEASE BURDEN

So, if routinely available surveillance data are severely limited, what other options exist to help us describe the disease burden?

Retrospective Surveys

Retrospective population-based, cross-sectional surveys have been employed in a number of countries to derive incidence estimates of AGE (however it is defined) (Table 2). Survey methods, case definitions, target populations, and periods of recall vary, although most of the recent studies have used telephone surveys to enquire about illness in the general population in the previous month. Incidence estimates range from 0.48 episode per person-year in a survey of children to 1.3 episodes per person-year in two general-population surveys in different parts of Canada. To overcome the problem of different case definitions and in an attempt to compare rates internationally, Scallan et al. developed uniform case definitions and reanalyzed data from four national, retrospective cross-sectional telephone surveys (6,087 respondents in Australia, 3,496 in Canada, 9,903 in Ireland, and 14,647 in the United States) (77). The prevalence of diarrhea among respondents in the 4 weeks before the interview was 7.6% in Canada, 7.6% in the United States, 6.4% in Australia, and 3.4% in Ireland. Despite the different health care systems in the four countries, similar proportions of respondents had sought medical care (approximately one in five). This work has led to the development of a standard case definition for surveys of this type (S. E. Majowicz, personal communication). Researchers are being encouraged to collect sufficient data to report prevalence in accordance with an international definition alongside any case definition developed locally for the purposes of answering the specific research question.

A major concern with retrospective studies is the potential for recall bias and for telescoping (9, 68). This is a situation in which subjects might remember an unpleasant illness as being more recent than it really was and, say, report illness that actually occurred more than a month ago as having happened within the last month. It is interesting that in the study by Lupo et al., where the period of recall was 1 week, the incidence estimate was lower (53). However, this survey included only children. An added limitation of retrospective surveys of self-reported illness is that no information on causative pathogens is gleaned. Set against their limitations is the fact that these types of survey are relatively inexpensive to perform and that serial surveys over time, conducted in the same way, can provide valuable trend information.

Prospective Studies

Table 3 summarizes a number of prospective studies that have been performed to determine the incidence of AGE and/or NoV infection. Once again the target populations, case definitions, sample sizes, and person-years of follow-up (where stated) differ between studies. Some of these studies

Table 2 Retrospective population-based cross-sectional survey studies of self-reported AGE

Yr(s) of study	Country	Method	Period of recall	Population surveyed	Sample size (completed usable interviews)	Incidence estimate episodes/person-yr	Reference
2005–2006	Canada (Ontario)	Telephone survey	28 days	All ages (English speaking)	2,090	1.17	75
2002–2003	Canada (British Columbia)	Telephone survey	28 days	All ages	4,612	1.3	85
2001–2002	Canada (Hamilton, Ontario)	Telephone survey	28 days	All ages	3,496	1.3	55
2001–2002	Australia	Telephone survey	4 wks	All ages	6,087	0.92	36
2000–2001	Ireland	Telephone survey	4 wks	All ages	9,903	0.6	77
1999–2000	Norway	Postal questionnaire	4 wks	All ages	1,843	1.2	49
1998–1999	United States	Telephone survey	4 wks	All ages	12,075	0.72	40
1996–2003	United States	Telephone surveys	4 wks	All ages	52,840	0.6 (avg) annual incidence	45
1984–1985	Italy	Telephone survey of physicians	1 wk	Children 0–12 yr old	8,164	0.48	53

Table 3 Incidence estimates for AGE or NoV infection from population-based prospective cohort studies

Yr(s) of study	Country	Type of population sample	Size of cohort	Person-yrs of follow-up	Incidence estimate per person-yr	Reference
2001–2002	United States (34 states)	Convenience sample of children aged 6 mo to 3 yrs presenting for well-child visits or non-diarrhea-related consultations	604	Not stated	2.21 (AGE)	91
1998–1999	The Netherlands	Random sample (all ages) of individuals registered with sentinel general practices from The Netherlands Institute of Primary Health Care	4,860	2,340	0.283 (AGE)	20
1993–1996	England	Random population sample (all ages) of individuals registered with a primary-care physician who was part of the MRC General Practice Research Framework	9,776	4,026	0.19 (AGE), 0.013 (NoV)	93
1989–1993	Brazil	All infants from two urban favelas from birth to 4 years of age	211	186	0.7 (NoV based on seroconversion)	83
Not stated	Mexico	Sample of students attending medical school	124	Not stated	0.36 (NoV based on seroconversion)	43
Not stated	Canada	Neonates and siblings in Winnipeg, Manitoba, and settlements in northern Canada	98 families	Not stated	0.15 (NoV)	34
1965–1971	United States	Random sample of households in the Tecumseh community	4,905 individuals	Not stated	1.2 (AGE)	59

have incorporated microbiological examination of stool samples. Vernacchio et al. collected baseline stool samples from a cohort of children recruited through a national research network of pediatricians and family practitioners and also collected symptomatic samples from children in the cohort who subsequently developed AGE during the 6-month period of follow-up (91). Compliance with sampling was 80.3% at baseline and 73.2% after an episode of diarrhea. The authors found that 4 (0.8%) of 484 samples were positive for NoV at baseline by RT-PCR compared with (1.9%) of 431 after an episode of diarrhea. In both England and The Netherlands, researchers determined the etiology of gastroenteritis in the community and in patients presenting to primary-care physicians. In England, 7% of cases (50 of 715) and 0.6% of controls (3 of 535) in the community cohort were positive for NoV by EM (86). In the case-control study of individuals presenting to primary-care physicians, 6.5% of cases (169 of 2,612) and 0.3% of controls (6 of 2,210) were positive for NoV. In The Netherlands, the corresponding proportions were 11%, 5.2%, 5.1%, and 1.1%, respectively (20, 21). A big methodologic difference between the two studies was the use of EM to detect NoV in the English study and the use of a more sensitive RT-PCR assay in the Dutch study. Moreover, the case definition in the Dutch study was stricter than that employed in the English study. In total, in England an estimated 9.4 million cases of IID occurred each year (one in five members of the population) and the most common etiologic agent was NoV (606,700 cases). In The Netherlands, there were an estimated 4.5 million cases, with the most common etiologic agent also being NoV.

Prospective cohort studies are very expensive to conduct, especially if there is an etiologic component to the study, and maintaining the participation of cohort members is essential. In general, prospective studies tend to produce lower incidence estimates than retrospective studies do. While telescoping is a concern in retrospective studies, it has been suggested that prospective studies with an etiologic component might, in fact, lead to underestimates of disease incidence. The hypothesis, as yet untested, is that requesting stool samples from individuals with symptoms of diarrhea and vomiting is a disincentive to reporting those symptoms in the first place. Despite these shortcomings, the fact that NoV is a major cause of AGE in developed countries is not in doubt.

Economic Assessments

Some researchers have attempted to determine the economic burden of AGE and/or NoV. In England, the cost of AGE was estimated at £743 ($1,492) million expressed in 1994-1995 prices. The cost per case of confirmed NoV infection was £176 ($353) (69). The total cost of AGE was

A\$342 (\$290) million in Australia and €345 (\$467) million in The Netherlands (38, 89). Although these studies were conducted at different times and using slightly different methods, the conclusions are clear: regardless of the fact that the majority of individual cases of AGE in developed countries are of mild to moderate severity, the overall costs of illness are substantial. Unfortunately, there are no economic data from developing countries.

ASSESSING THE PROPORTION OF VIRAL GASTROENTERITIS THAT IS FOOD BORNE

Having developed an estimate for NoV in the community, the next big challenge is to try to estimate the proportion that is likely to be food borne, as opposed to transmitted via other routes (e.g., person-to-person transmission). Table 4 illustrates the very wide variation in both estimated numbers of cases of NoV in the population and the proportion judged to be transmitted through food. The various methods that have been used to assess the proportion of viral gastroenteritis that is food borne are summarized below.

Analytical Epidemiologic Studies

Ideally, there would be data from well-conducted analytical epidemiologic studies that have considered transmission routes as the primary outcome. Such studies are extremely sparse for sporadic NoV. In a case-control study nested within a prospective cohort study in The Netherlands, independent risk factors for NoV gastroenteritis were (i) having a household member with gastroenteritis, (ii) having contact with a person with gastroenteritis outside the household, and (iii) using poor food-handling hygiene. The population-attributable risk fractions were 17%, 56%, and 47%, respectively (23). Around 12% of NoV infections were attributed to food-borne transmission in this study. However, in a case-control study in Switzerland, Fretz et al. failed to identify any associations between illness and consumption of food (28). They concluded that person-to-person transmission was the most important route of transmission for community-acquired, sporadic NoV infection, since 39% of all patients reported having had contact with ill persons before their illness. Furthermore, since 33% of patients reported having had contact with ill persons, mainly within family groups, after their own illness, the authors concluded that a substantial proportion of patients were part of family mini-outbreaks.

So how might the apparent inconsistencies between these findings be explained? Clearly, differences in study design might account for the conflicting results. One of the persistent problems with case-control studies of

Table 4 Estimating the burden of NoV that is food borne

| Period of study | Country | Population-based AGE estimate available? | Estimated total no. of NoV cases | Estimated no. (proportion) that were food borne | Methods for attributing food borne transmission | | | Reference |
					Analytical epidemiologic studies?	Expert opinion?	Outbreak surveillance data?	
1990s	United States	Yes (FoodNet retrospective telephone surveys)	23,000,000	9,200,000 (40%)	Yes	Yes	Yes	56
1990s	France	No	176,500	70,600 (40%)	No	Yes	No[a]	88
1995	England and Wales	Yes (Prospective cohort study)	647,701	69,628 (11%)	Yes	No	Yes	1
1999	The Netherlands	Yes (Prospective cohort study)	650,000	80,000 (12%)	Yes	No	No	23
2000	Australia	Yes (retrospective telephone survey)	1,832,000	446,000 (25%)	Yes	Yes	Yes	35

[a] Estimates used were based on those from The Netherlands.

sporadic infection is the immune status of controls. Few investigators attempt to identify what proportion of their controls might have been immune to the infection in question and therefore might have misclassified their controls. Including immune controls in risk estimates will serve to dilute the association between exposure and disease.

Similarly, during food-borne outbreaks, secondary person-to-person transmission is common. However, because the incubation period can be very short, it can be difficult to separate primary from secondary cases since early onset of illness in secondary cases often overlaps with late onset in primary cases. Gotz et al. took the opportunity to study secondary transmission into households following a large food-borne outbreak of NoV in the community (32). They found that exposure to vegetable salad was associated clearly with the patients who had an onset of symptoms on days 1 to 3 only. The association disappeared for patients who had an onset of symptoms on day 4 or later. The investigators estimated that the median incubation time of food-borne infection was 34 h (range, 30 to 43 h). Misclassification of secondary cases of infection as primary cases in studies of sporadic disease is therefore a distinct possibility and would serve to dilute any associations between food exposure and development of disease.

Expert Opinion

Where good epidemiologic or surveillance data are unavailable, inadequate or highly dubious expert opinion is often called on to fill the gap (6). A commonly used method for eliciting expert opinion is the Delphi technique (18). It assumes that pooled intelligence enhances individual judgment and captures the collective opinion of a group of experts. The conventional Delphi technique employs a series of questionnaires to generate expert opinion made anonymous over a number of rounds. Critical to its validity are questionnaire development, the definition of consensus and how to interpret disagreement, the criteria for and selection of the expert panel, the sample size, and the approach to data analysis. In the context of food-borne disease, expert opinion is likely to be based on experience, knowledge gleaned from outbreak investigations, and the peer-reviewed literature. Yet what appears in outbreak surveillance systems and in the peer-reviewed literature will be biased by the vigor with which NoV outbreak investigations are pursued (61). Furthermore, in a review of 1,763 outbreaks of food-borne disease in England and Wales, in which food vehicles reported to a systematic surveillance system were compared with those published in the peer-reviewed literature, publications in the peer-reviewed literature favored unusual food vehicles or novel events. This is not entirely surprising, given the mission of peer-reviewed journals, but might also influence expert judgments (62). The use of expert opinion is limited by the fact that it is based largely on perception rather than hard data (6).

Outbreak Data

Many of the investigations shown in Table 4 made use of outbreak data to estimate the proportion of food-borne NoV. Extrapolating information from outbreak data sets in an attempt to describe the food-borne NoV burden is not straightforward. The biggest risk is that outbreak cases might not be representative of all cases in the population. However, for an outbreak-related disease, this might be slightly less problematic than for some other pathogens, for example *Campylobacter* infection, where fewer than 1% of cases occur as part of outbreaks. Indeed, it is noteworthy that the estimate by Adak et al. (11%), who used outbreak data to determine the proportion of NoV that is food borne, is closer to that by de Wit et al. (12%), who employed a case-control study, than either is to the estimate by Mead et al. (40%) (1, 23, 56). Further support for estimates closer to that of Adak et al. and de Wit et al. comes from a recently published review of outbreaks of NoV in Switzerland, which reported that 13% of outbreaks were food borne (29). Had Mead et al. used a more conservative estimate for the proportion of food-borne NoV infections, the number of cases would have dropped from 9.2 million to 2.1 million, and 76 million cases of food-borne disease in the United States annually would have become 69 million. However, Widdowson et al. have produced estimates that support the work by Mead (94). Clearly, the proportion that is chosen is affected enormously by the surveillance system from which the data are drawn and, in turn, affects hugely the perceived total burden of food-borne NoV and food-borne disease as a whole.

FOOD ATTRIBUTION

Few investigators have attempted to determine the NoV burden in contaminated foods. This last step is essential to guide primary prevention programs targeted along the food chain but is also extremely difficult. In an analysis by Adak et al., the highest disease risk from any food group was from shellfish, at 646 cases/10^6 servings (2). Preharvest contamination of oysters with NoV had a major impact in generating cases of disease. In addition in this study, the impact of infected food handlers was examined. Over 67,000 cases were attributed to infected food handlers, and the pathogen most frequently transmitted by infected food handlers was NoV. These analyses relied heavily on outbreak surveillance data, drawing on 766 outbreaks in which a single vehicle of infection was identified; 612 outbreaks that were reported as food borne but in which no food vehicle was implicated were excluded from analysis. The assumption was that distribution of foods in the subset of outbreaks in which a vehicle was identified was representative of the complete population of outbreaks. However, certain vehicles may be more likely to be implicated in outbreak investigations than others, especially if investigators preferentially collect data on the types of food that are perceived as high risk or when laboratory methods vary in sensitivity according to food type. Therefore, a systematic vehicle detection bias might lead to an underestimation of the contribution and risks attributable to foods that were rarely implicated in outbreak investigations, e.g., salad items, fruit (implicated in NoV outbreaks), and background ingredients such as herbs and spices. This vehicle detection bias operates on top of the outbreak detection and investigation biases referred to earlier.

Hoffmann et al. chose expert elicitation to link food-borne illnesses to food commodities in the United States (39). They assembled 42 experts to attribute U.S. food-borne illnesses caused by the nine pathogens under surveillance in FoodNet, *Toxoplasma gondii*, and NoVs to consumption of foods in 11 broad categories. For each pathogen they asked each expert to give his or her best estimate of the distribution of food-borne illnesses associated with each of the food categories and the 90% confidence bounds on each of their estimates. They found that 3.4 million of the 9.2 million cases of NoV infection in the United States each year were attributed to contaminated produce (salad vegetables, etc.) and that 3.1 million were attributed to contaminated seafood. Indeed, NoV-contaminated produce and seafood ranked in the top 5 pathogen-food pairs for illnesses and admissions to hospital but were not in the top 10 pathogen-food pairs for deaths.

IMPROVING THE ASSESSMENT OF THE BURDEN OF FOOD-BORNE DISEASE THAT IS DUE TO VIRUSES

Given all the practical difficulties in trying to estimate the burden of food-borne NoV, and the biases that exist in all forms of surveillance, most projections about disease burden are likely to be inaccurate at best and positively misleading at worst. Few countries have undertaken disease burden estimates that include etiology. So how can we improve our assessment of food-borne viral gastroenteritis?

Population Surveys
The development of a standardized case definition for population surveys of AGE is to be welcomed. This will enhance the ability to undertake interna-

tional comparisons. Adding an etiology component to population surveys should be encouraged wherever possible: perhaps the inclusion of salivary antibody assays in population-based prospective studies of endemic NoV infection (58). Collecting saliva samples at 2-week intervals captures the rapid rise and fall of NoV-specific salivary antibody titers and could be used to estimate the incidence of NoV infection in longitudinal studies. It is far easier than collecting stool or serum specimens, is more convenient, and is associated with fewer refusals from subjects asked to provide specimens. Where stool samples can be obtained, lab-on-a-chip technology that allows simultaneous detection and genotyping of NoVs is suitable for use in large-scale epidemiologic studies for both diagnosis and virus typing (92), although frequent changes in the panel of detection primers may be required (41).

Disease burden data from developing countries are relatively sparse. Several seroprevalence surveys have been conducted in developing countries including Brazil (83), Mexico (42, 66), South Africa (79), and Bangladesh (7). They show higher age-specific NoV seroprevalence in young children compared with their counterparts in developed countries. These studies have depended on blood samples, but salivary assays may also be useful tools for studies in developing countries and for pediatric or difficult-to-reach populations (58). It should be borne in mind that the risk factors for NoV infection in developing countries may be different, e.g., exposure to sewage-contaminated water. However, where contaminated water is used to irrigate food crops, food-borne transmission of NoV occurs, as was demonstrated in Chile (65); thus, food-borne transmission should not be overlooked.

Finally, incorporating assessments of immunity among controls in population-based case-control studies should lead to more accurate determination of risk factors for sporadic infection.

Investigation of Diarrheal Disease

A key to improving the recognition of NoV infection, either sporadic disease or outbreaks, depends on implementing state-of-the-art diagnostics. Although RT-PCR assays have revolutionized the detection of NoV in clinical specimens, they are costly, not routinely used in clinical laboratories, and not particularly suitable for processing large numbers of samples (92). It is essential that we improve our understanding of the contribution of NoV and other viruses to the burden of AGE by using DNA microarray technology in diagnostic laboratories, perhaps as part of national sentinel networks. Similarly, employing these methods in outbreak situations, thereby reducing the time needed to recognize an outbreak, might help to uncover a food-borne event seeding subsequent person-to-person transmission.

Making Better Use of Existing Sources of Surveillance Data

A major step forward would be to ensure that more food-borne disease surveillance programs include NoV as one of the target pathogens for surveillance. Integrating virological investigations with standardized epidemiologic data should enhance the detection of NoV and hence our understanding of the role of NoV in food-borne disease.

As far as the systems that are in place at the moment, none adequately describes the burden of viral gastroenteritis that is transmitted through food. Each sheds chinks of light on different parts of the picture. This means developing new epidemiologic techniques for combining and analyzing data from different sources, making use of sensitivity analyses and weighting to correct for the biases inherent in each system (as far as this is possible). Incorporating data from new data sources such as pharmacy sales and clinical telephone triage systems, alongside more traditional communicable-disease surveillance systems, affords the opportunity to employ techniques like capture-recapture. Despite the technological advances that have taken place in food microbiology, routine viral monitoring of food commodities in the foreseeable future is unlikely (24).

CONCLUSION

Few would argue that food-borne viruses are unimportant. It is clear, however, that our current assessment of their impact is wrong. It is likely that, at every turn, we underestimate their significance. So what we do not yet know is just how wrong we are!

REFERENCES

1. Adak, G. K., S. M. Long, and S. J. O'Brien. 2002. Trends in indigenous foodborne disease and deaths, England and Wales: 1992 to 2000. *Gut* **51**:832–841.
2. Adak, G. K., S. M. Meakins, H. Yip, B. A. Lopman, and S. J. O'Brien. 2005. Disease risks from foods, England and Wales, 1996–2000. *Emerg. Infect. Dis.* **11**:365–372.
3. Amar, C. F., C. L. East, J. Gray, M. Iturriza-Gomara, E. A. Maclure, and J. McLauchlin. 2007. Detection by PCR of eight groups of enteric pathogens in 4,627 faecal samples: re-examination of the English case-control Infectious Intestinal Disease Study (1993–1996). *Eur. J. Clin. Microbiol. Infect. Dis.* **26**:311–323.
4. Ashbolt, R., R. Givney, J. E. Gregory, G. Hall, R. Hundy, M. Kirk, I. McKay, L. Meuleners, G. Millard, J. Raupach, P. Roche, N. Prasopa-Plaizier, M. K. Sama, R. Stafford, N. Tomaska, L. Unicomb, and C. Williams for the OzFoodNet Working Group. 2002. Enhancing food-borne disease surveillance across Australia in 2001: the OzFoodNet Working Group. *Commun. Dis. Intell.* **26**:375–406.
5. Balter, S., D. Weiss, H. Hanson, V. Reddy, D. Das, and R Heffernan. 2005. Three years of emergency department gastrointestinal syndromic surveillance in New York City: what have we found? *Morb. Mortal. Wkly. Rep.* **54**(Suppl.):175–180.

6. Batz, M. B., M. P. Doyle, G. Morris, Jr., J. Painter, R. Singh, R. V. Tauxe, M. R. Taylor, and D. M. Lo Fo Wong for the Food Attribution Working Group. 2005. Attributing illness to food. *Emerg. Infect. Dis.* **11**:993–999.

7. Black, R. E., H. B. Greenberg, A. Z. Kapikian, K. H. Brown, and S. Becker. 1982. Acquisition of serum antibody to Norwalk virus and rotavirus and relation to diarrhea in a longitudinal study of young children in rural Bangladesh. *J. Infect. Dis.* **145**:483–489.

8. Blanton, L. H., S. M. Adams, R. S. Beard, G. Wei, S. N. Bulens, M. A. Widdowson, R. I. Glass, and S. S. Monroe. 2006. Molecular and epidemiologic trends of caliciviruses associated with outbreaks of acute gastroenteritis in the United States, 2000–2004. *J. Infect. Dis.* **193**:413–421.

9. Bruijnzeels, M. A., J. C. van der Wouden, M. Foets, A. Prins, and W. J. van den Heuvel. 1998. Validity and accuracy of interview and diary data on children's medical utilisation in The Netherlands. *J. Epidemiol. Community Health* **52**:65–69.

10. Buesa, J., B. Collado, P. Lopez-Andujar, R. Abu-Mallouh, J. Rodriguez Diaz, A. Garcia Diaz, J. Prat, S. Guix, T. Llovet, G. Prats, and A. Bosch. 2002. Molecular epidemiology of caliciviruses causing outbreaks and sporadic cases of acute gastroenteritis in Spain. *J. Clin. Microbiol.* **40**:2854–2859.

11. Butot, S., T. Putallaz, and G. Sánchez. 2007. Procedure for rapid concentration and detection of enteric viruses from berries and vegetables. *Appl. Environ. Microbiol.* **73**:186–192.

12. Colomba, C., S. De Grazia, G. M. Giammanco, L. Saporito, F. Scarlata, L. Titone, and S. Arista. 2006. Viral gastroenteritis in children hospitalised in Sicily, Italy. *Eur. J. Clin. Microbiol. Infect. Dis.* **25**:570–575.

13. Cooper, D., E. Arnold, G. Smith, V. Hollyoak, F. Chinemana, M. Baker, and S. J. O'Brien. 2005. The effect of deprivation, age and sex on NHS Direct call rates. *Br. J. Gen. Pract.* **55**:287–291.

14. Cooper, D. L., G. E. Smith, S. J. O'Brien, V. A. Hollyoak, and M. Baker. 2003. What can analysis of calls to NHS direct tell us about the epidemiology of gastrointestinal infections in the community? *J. Infect.* **46**:101–105.

15. Cooper, D. L., N. Q. Verlander, G. E. Smith, A. Charlett, E. Gerard, L. Willocks, and S. O'Brien. 2006. Can syndromic surveillance data detect local outbreaks of communicable disease? A model using a historical cryptosporidiosis outbreak. *Epidemiol. Infect.* **134**:13–20.

16. Dai, Y. C., J. Nie, Y. Liu, Y. M. Yao, Z. F. Li, and S. Y. Yu. 2004. Preliminary study of human calicivirus infection in Guangzhou. *Di Yi Jun Yi Da Xue Xue Bao* **24**:296–299.

17. Derby, M. P., J. McNally, J. Ranger-Moore, L. Hulette, R. Villar, T. Hysong, E. MacNeill, M. Lebowitz, and J. Burgess. 2005. Poison Control Center-based syndromic surveillance for food-borne illness. *Morb. Mortal. Wkly. Rep.* **54**(Suppl.):35–40.

18. de Villiers, M. R., P. J. de Villiers, and A. P. Kent. 2005. The Delphi technique in health sciences education research. *Med. Teach.* **27**:639–643.

19. de Wit, M. A., L. M. Kortbeek, M. P. Koopmans, C. J. de Jager, W. J. Wannet, A. I. Bartelds, and Y. T. van Duynhoven. 2001. A comparison of gastroenteritis in a general practice-based study and a community-based study. *Epidemiol. Infect.* **127**:389–397.

20. de Wit, M. A., M. P. Koopmans, L. M. Kortbeek, W. J. Wannet, J. Vinjé, F. van Leusden, A. I. Bartelds, and Y. T. van Duynhoven. 2001. Sensor, a population-based cohort study

on gastroenteritis in the Netherlands: incidence and etiology. *Am. J. Epidemiol.* **154**:666–674.

21. de Wit, M. A., M. P. Koopmans, L. M. Kortbeek, N. J. van Leeuwen, J. Vinjé and Y. T. van Duynhoven. 2001. Etiology of gastroenteritis in sentinel general practices in The Netherlands. *Clin. Infect. Dis.* **33**:280–288.

22. de Wit, M. A., M. P. Koopmans, L. M. Kortbeek, N. J. van Leeuwen, A. I. Bartelds, and Y. T. van Duynhoven. 2001. Gastroenteritis in sentinel general practices, The Netherlands. *Emerg. Infect. Dis.* **7**:82–91.

23. de Wit, M. A., M. P. Koopmans, and Y. T. van Duynhoven. 2003. Risk factors for norovirus, Sapporo-like virus, and group A rotavirus gastroenteritis. *Emerg. Infect. Dis.* **9**:1563–1570.

24. Duizer, E., and M. Koopmans. 2006. Tracking emerging pathogens: the case of noroviruses, p. 77–110. *In* Y. Motarjemi and M. Adams (ed.), *Emerging Food-Borne Pathogens*. Woodhead Publishing, Cambridge, United Kingdom.

25. Flahault, A., T. Blanchon, Y. Dorleans, L. Toubiana, J. F. Vibert, and A. J. Valleron. 2006. Virtual surveillance of communicable diseases: a 20-year experience in France. *Stat. Methods Med. Res.* **15**:413–421.

26. Foley, B., J. O'Mahony, S. M. Morgan, C. Hill, and J. G. Morgan. 2000. Detection of sporadic cases of Norwalk-like virus (NLV) and astrovirus infection in a single Irish hospital from 1996 to 1998. *J. Clin. Virol.* **17**:109–117.

27. Fretz, R., L. Herrmann, A. Christen, P. Svoboda, O. Dubuis, E. H. Viollier, M. Tanner, and A. Baumgartner. 2005. Frequency of Norovirus in stool samples from patients with gastrointestinal symptoms in Switzerland. *Eur. J. Clin. Microbiol. Infect. Dis.* **24**:214–216.

28. Fretz, R., P. Svoboda, D. Schorr, M. Tanner, and A. Baumgartner. 2005. Risk factors for infections with Norovirus gastrointestinal illness in Switzerland. *Eur. J. Clin. Microbiol. Infect. Dis.* **24**:256–261.

29. Fretz, R., P. Svoboda, T. M. Luthi, M. Tanner, and A. Baumgartner. 2005. Outbreaks of gastroenteritis due to infections with Norovirus in Switzerland, 2001–2003. *Epidemiol. Infect.* **133**:429–437.

30. Froggatt, P. C., I. B. Vipond, C. R. Ashley, P. R. Lambden, I. N. Clarke, and E. O. Caul. 2004. Surveillance of norovirus infection in a study of sporadic childhood gastroenteritis in South West England and South Wales, during one winter season (1999–2000). *J. Med. Virol.* **72**:307–311.

31. Frosst, G. O., S. E. Majowicz, and V. L. Edge. 2006. Factors associated with the use of over-the-counter medications in cases of acute gastroenteritis in Hamilton, Ontario. *Can. J. Public Health* **97**:489–493.

32. Gotz, H., K. Ekdahl, J. Lindback, B. de Jong, K. O. Hedlund, and J. Giesecke. 2001. Clinical spectrum and transmission characteristics of infection with Norwalk-like virus: findings from a large community outbreak in Sweden. *Clin. Infect. Dis.* **33**:622–628.

33. Guévremont, E., J. Brassard, A. Houde, C. Simard, and Y. L. Trottier. 2006. Development of an extraction and concentration procedure and comparison of RT-PCR primer systems for the detection of hepatitis A virus and norovirus GII in green onions. *J. Virol. Methods* **134**:130–135.

34. Gurwith, M., W. Wenman, D. Gurwith, J. Brunton, S. Feltham, and H. Greenberg. 1983. Diarrhea among infants and young children in Canada: a longitudinal study in three northern communities. *J. Infect. Dis.* **147**:685–692.

35. Hall, G., M. D. Kirk, N. Becker, J. E. Gregory, L. Unicomb, G. Millard, R. Stafford, and K. Lalor for the OzFoodNet Working Group. 2005. Estimating food-borne gastroenteritis, Australia. *Emerg. Infect. Dis.* **11**:1257–1264.

36. Hall, G. V., M. D. Kirk, R. Ashbolt, R. Stafford, and K. Lalor. 2006. Frequency of infectious gastrointestinal illness in Australia, 2002: regional, seasonal and demographic variation. *Epidemiol. Infect.* **134**:111–118.

37. Hansman, G. S., L. T. Doan, T. A. Kguyen, S. Okitsu, K. Katayama, S. Ogawa, K. Natori, N. Takeda, Y. Kato, O. Nishio, M. Noda, and H. Ushijima. 2004. Detection of norovirus and sapovirus infection among children with gastroenteritis in Ho Chi Minh City, Vietnam. *Arch. Virol.* **149**:1673–1688.

38. Hellard, M. E., M. I. Sinclair, A. H. Harris, M. Kirk, and C. K. Fairley. 2003. Cost of community gastroenteritis. *J. Gastroenterol. Hepatol.* **18**:322–328.

39. Hoffmann, S., P. Fischbeck, A. Krupnick, and M. McWilliams. 2007. Using expert elicitation to link food-borne illnesses in the United States to foods. *J. Food Prot.* **70**:1220–1229.

40. Imhoff, B., D. Morse, B. Shiferaw, M. Hawkins, D. Vugia, S. Lance-Parker, J. Hadler, C. Medus, M. Kennedy, M. R. Moore, and T. Van Gilder for the Emerging Infections Program FoodNet Working Group. 2004. Burden of self-reported acute diarrheal illness in FoodNet surveillance areas, 1998–1999. *Clin. Infect. Dis.* **38**(Suppl. 3):**S219–S226**.

41. Jaaskelainen, A. J., and L. Maunula. 2006. Applicability of microarray technique for the detection of noro- and astroviruses. *J. Virol. Methods* **136**:210–216.

42. Jiang, X., D. O. Matson, F. R. Velazquez, J. J. Calva, W. M. Zhong, J. Hu, G. M. Ruiz-Palacios, and L. K. Pickering. 1995. Study of Norwalk-related viruses in Mexican children. *J. Med. Virol.* **47**:309–316.

43. Johnson, P. C., J. Hoy, J. J. Mathewson, C. D. Ericsson, and H. L. DuPont. 1990. Occurrence of Norwalk virus infections among adults in Mexico. *J. Infect. Dis.* **162**:389–393.

44. Jones, N. F., and R. Marshall. 2004. Evaluation of an electronic general-practitioner-based syndromic surveillance system—Auckland, New Zealand, 2000–2001. *Morb. Mortal. Wkly. Rep.* **53**(Suppl.):173–178.

45. Jones, T. F., M. B. McMillian, E. Scallan, P. D. Frenzen, A. B. Cronquist, S. Thomas, and F. J. Angulo. 2007. A population-based estimate of the substantial burden of diarrheal disease in the United States; FoodNet, 1996–2003. *Epidemiol. Infect.* **135**:293–301.

46. Jones, T. F., E. Scallan, and F. J. Angulo. 2007. FoodNet: overview of a decade of achievement. *Food-Borne Pathog. Dis.* **4**:60–66.

47. Kaplan, J. E., G. W. Gary, R. C. Baron, N. Singh, L. B. Schonberger, R. Feldman, and H. B. Greenberg. 1982. Epidemiology of Norwalk gastroenteritis and the role of Norwalk virus in outbreaks of acute nonbacterial gastroenteritis. *Ann. Intern. Med.* **96**:756–761.

48. Kirkwood, C. D., R. Clark, N. Bogdanonic-Sakran, and R. F. Bishop. 2005. A 5-year study of the prevalence and genetic diversity of human caliciviruses associated with sporadic cases of acute gastroenteritis in young children admitted to hospital in Melbourne, Australia (1998–2002). *J. Med. Virol.* **77**:96–101.

49. Kuusi, M., P. Aavitsland, B. Gondrosen, and G. Kapperud. 2003. Incidence of gastroenteritis in Norway—a population-based survey. *Epidemiol. Infect.* **131**:591–597.

50. Lau, C. S., D. A. Wong, L. K. Tong, J. Y. Lo, A. M. Ma, P. K. Cheng, and W. W. Lim. 2004. High rate and changing molecular epidemiology pattern of norovirus infections in sporadic cases and outbreaks of gastroenteritis in Hong Kong. *J. Med. Virol.* 73:113–117.

51. Leggitt, P. R., and L. A. Jaykus. 2000. Detection methods for human enteric viruses in representative foods. *J. Food Prot.* 63:1738–1744.

52. Lopman, B. A., M. H. Reacher, Y, Van Duijnhoven, F. X. Hanon, D. Brown, and M. Koopmans. 2003. Viral gastroenteritis outbreaks in Europe, 1995–2000. *Emerg. Infect. Dis.* 9:90–96.

53. Lupo, L., V. De Grandi, E. Ganci, A, Nastri, S. Ielo, A. Mistretta, and G. Giammanco. 1989. Epidemiology of acute diarrhea in children living in Sicily. *Eur. J. Epidemiol.* 5:400–402.

54. MacDougall, L., S. Majowicz, K. Dore, J. Flint, K. Thomas, S. Kovacs, and P. Sockett. 2007. Under-reporting of infectious gastrointestinal illness in British Columbia, Canada: who is counted in provincial communicable disease statistics? *Epidemiol. Infect.* Apr. 16:1–9 [Epub ahead of print].

55. Majowicz, S. E., K. Dore, J. A. Flint, V. L. Edge, S. Read, M. C. Buffett, S. McEwen, W. B. McNab, D. Stacey, P. Sockett, and J. B. Wilson. 2004. Magnitude and distribution of acute, self-reported gastrointestinal illness in a Canadian community. *Epidemiol. Infect.* 132:607–617.

56. Mead, P. S., L. Slutsker, V. Dietz, L. F. McCaig, J. S. Bresee, C. Shapiro, P. M. Griffin, and R. V. Tauxe. 1999. Food-related illness and death in the United States. *Emerg. Infect. Dis.* 5:607–625.

57. Medici, M. C., M. Martinelli, L. A. Abelli, F. M. Ruggeri, I. Di Bartolo, M. C. Arcangeletti, F. Pinardi, F. De Conto, G. Izzi, S. Bernasconi, C. Chezzi, and G. Dettori. 2006. Molecular epidemiology of norovirus infections in sporadic cases of viral gastroenteritis among children in Northern Italy. *J. Med. Virol.* 78:1486–1492.

58. Moe, C. L., A. Sair, L. Lindesmith, M. K. Estes, and L. A. Jaykus. 2004. Diagnosis of Norwalk virus infection by indirect enzyme immunoassay detection of salivary antibodies to recombinant Norwalk virus antigen. *Clin. Diagn. Lab. Immunol.* 11:1028–1034.

59. Monto, A. S., and J. S. Koopman. 1980. The Tecumseh study. XI. Occurrence of acute enteric illness in the community. *Am. J. Epidemiol.* 112:323–333.

60. Nebot, M., and M. A. Llauger. 1992. Self care of common health disorders: results of a telephone survey of the general population. *Med. Clin. (Barcelona)* 99:420–424.

61. O'Brien, S. J., R. Elson, I. A. Gillespie, G. K. Adak, and J. M. Cowden. 2002. Surveillance of food-borne outbreaks of infectious intestinal disease in England and Wales 1992–1999: contributing to evidence-based food policy? *Public Health* 116:75–80.

62. O'Brien, S. J., I. A. Gillespie, M. A. Sivanesan, R. Elson, C. Hughes, and G. K. Adak. 2006. Publication bias in food-borne outbreaks of infectious intestinal disease and its implications for evidence-based food policy. England and Wales 1992–2003. *Epidemiol. Infect.* 134:667–674.

63. O'Brien, S. J., and I. S. T. Fisher. 2006. Surveillance of emerging pathogen in Europe, p. 50–76. *In* Y. Motarjemi and M. Adams (ed.), *Emerging Food-Borne Pathogens.* Woodhead Publishing, Cambridge, United Kindgom.

64. Okada, M., T. Ogawa, I. Kaiho, and K. Shinozaki. 2005. Genetic analysis of noroviruses in Chiba Prefecture, Japan, between 1999 and 2004. *J. Clin. Microbiol.* 43:439–401.

65. O'Ryan, M. L., P. A. Vial, N. Mamani, X. Jiang, M. K. Estes, C. Ferrecio, H. Lakkis, and D. O. Matson. 1998. Seroprevalence of Norwalk virus and Mexico virus in Chilean individuals: assessment of independent risk factors for antibody acquisition. *Clin. Infect. Dis.* **27**:789–795.

66. Peasey, A. E., G. M. Ruiz-Palacios, M. Quigley, W. Newsholme, J. Martinez, G. Rosales, X. Jiang, and U. J. Blumenthal. 2004. Seroepidemiology and risk factors for sporadic norovirus/Mexico strain. *J. Infect. Dis.* **189**:2027–2036.

67. Phan, T. G., F. Yagyu, V. Kozlov, A. Kozlov, S. Okitsu, W. E. Muller, and H. Ushijima. 2006. Viral gastroenteritis and genetic characterization of recombinant norovirus circulating in Eastern Russia. *Clin. Lab.* **52**:247–253.

68. Pickles, A., M. Neale, E. Simonoff, M. Rutter, J. Hewitt, J. Meyer, R. Crouchley, J. Silberg, and L. Eaves. 1994. A simple method for censored age-of-onset data subject to recall bias: mothers' reports of age of puberty in male twins. *Behav. Genet.* **24**:457–468.

69. Roberts, J. A., P. Cumberland, P. N. Sockett, J. Wheeler, L. C. Rodrigues, D. Sethi, and P. J. Roderick for the Infectious Intestinal Disease Study Executive. 2003. The study of infectious intestinal disease in England: socio-economic impact. *Epidemiol. Infect.* **130**: 1–11.

70. Rockx, B., M. De Wit, H. Vennema, J. Vinje, E. De Bruin, Y. Van Duynhoven, and M. Koopmans. 2002. Natural history of human calicivirus infection: a prospective cohort study. *Clin. Infect. Dis.* **35**:246–253.

71. Rooney, R., S. J. O'Brien, R. Mitchell, R. Stanwell-Smith, and P. E. Cook. 2000. Survey of local authority approaches to investigating sporadic cases of suspected food poisoning. *Commun. Dis. Public Health* **3**:101–105

72. Sagalova, O. I., A. T. Podkolzin, N. I. Abramycheva, O. A. Pishchulova, and V. V. Maleev. 2006. Acute intestinal infections of virus etiology in adults. *Ter. Arkh.* **78**:17–23.

73. Sair, A. I., D. H. D'Souza, C. L. Moe, and L. A. Jaykus. 2002. Improved detection of human enteric viruses in foods by RT-PCR. *J. Virol. Methods* **100**:57–69.

74. Sarfati, D., M. N. Bates, N. Garrett, and M. G. Baker. 1997. Acute gastroenteritis diagnostic practices of New Zealand general practitioners. *N. Z. Med. J.* **110**:354–356.

75. Sargeant, J. M., S. E. Majowicz, and J. Snelgrove. 2007. The burden of acute gastrointestinal illness in Ontario, Canada, 2005–2006. *Epidemiol. Infect.* Jun 13:1–10 [Epub ahead of print].

76. Scallan, E., T. F. Jones, A. Cronquist, S. Thomas, P. Frenzen, D. Hoefer, C. Medus, and F. J. Angulo for the FoodNet Working Group. 2006. Factors associated with seeking medical care and submitting a stool sample in estimating the burden of food-borne illness. *Food-Borne Pathog. Dis.* **3**:432–438.

77. Scallan, E., S. E. Majowicz, G. Hall, A. Banerjee, C. L. Bowman, L. Daly, T. Jones, M. D. Kirk, M. Fitzgerald, and F. J. Angulo. 2005. Prevalence of diarrhoea in the community in Australia, Canada, Ireland, and the United States. *Int. J. Epidemiol.* **34**:454–460.

78. Sinclair, M. I., M. E. Hellard, R. Wolfe, T. Z. Mitakakis, K. Leder, and C. K. Fairley. 2005. Pathogens causing community gastroenteritis in Australia. *J. Gastroenterol. Hepatol.* **20**:1685–1690.

79. Smit, T. K., P. Bos, I. Peenze, X. Jiang, M. K. Estes, and A. D. Steele. 1999. Seroepidemiological study of genogroup I and II calicivirus infections in South and southern Africa. *J. Med. Virol.* **59**:227–331.

80. Steiner-Sichel, L., J. Greenko, R. Heffernan, M. Layton, and D. Weiss. 2004. Field investigations of emergency department syndromic surveillance signals—New York City. *Morb. Mortal. Wkly. Rep.* **53**(Suppl.):184–189.

81. Subekti, D., M. Lesmana, P. Tjaniadi, N. Safari, E. Frazier, C. Simanjuntak, S. Komalarini, J. Taslim, J. R. Campbell, and B. A. Oyofo. 2002. Incidence of Norwalk-like viruses, rotavirus and adenovirus infection in patients with acute gastroenteritis in Jakarta, Indonesia. *FEMS Immunol. Med. Microbiol.* **33**:27–33.

82. Sumi, A., N. Kobayashi, and N. Ohtomo. 2005. Proportion of sporadic gastroenteritis cases caused by rotavirus, norovirus, adenovirus and bacteria in Japan from January 2000 to December 2003. *Microbiol. Immunol.* **49**:745–756.

83. Talal, A. H., C. L. Moe, A. A. Lima, K. A. Weigle, L. Barrett, S. I. Bangdiwala, M. K. Estes, and R. L. Guerrant. 2000. Seroprevalence and seroincidence of Norwalk-like virus infection among Brazilian infants and children. *J. Med. Virol.* **61**:117–124.

84. Tam, C. C., L. C. Rodrigues, and S. J. O'Brien. 2003. The study of infectious intestinal disease in England: what risk factors for presentation to general practice tell us about potential for selection bias in case-control studies of reported cases of diarrhoea. *Int. J. Epidemiol.* **32**:99–105.

85. Thomas, M. K., S. E. Majowicz, L. MacDougall, P. N. Sockett, S. J. Kovacs, M. Fyfe, V. L. Edge, K. Dore, J. A. Flint, S. Henson, and A. Q. Jones. 2006. Population distribution and burden of acute gastrointestinal illness in British Columbia, Canada. *BMC Public Health* **6**:307.

86. Tompkins, D. S., M. J. Hudson, H. R. Smith, R. P. Eglin, J. G. Wheeler, M. M. Brett, R. J. Owen, J. S. Brazier, P. Cumberland, V. King, and P. E. Cook. 1999. A study of infectious intestinal disease in England: microbiological findings in cases and controls. *Commun. Dis. Public Health* **2**:108–113.

87. Turcios, R. M., M. A. Widdowson, A. C. Sulka, P. S. Mead, and R. I. Glass. 2006. Reevaluation of epidemiological criteria for identifying outbreaks of acute gastroenteritis due to norovirus: United States, 1998–2000. *Clin. Infect. Dis.* **42**:964–969.

88. Vaillant, V., H. de Valk, E. Baron, T. Ancelle, P. Colin, M. C. Delmas, B. Dufour, R. Pouillot, Y. Le Strat, P. Weinbreck, E. Jougla, and J. C. Desenclos. 2005. Food-borne infections in France. *Food-Borne Pathog. Dis.* **2**:221–232.

89. van den Brandhof, W. E., G. A. de Wit, M. A. de Wit, and Y. T. van Duynhoven. 2004. Costs of gastroenteritis in The Netherlands. *Epidemiol. Infect.* **132**:211–221.

90. van den Brandhof, W. E., A. I. Bartelds, M. P. Koopmans, and Y. T. van Duynhoven. 2006. General practitioner practices in requesting laboratory tests for patients with gastroenteritis in The Netherlands, 2001–2002. *BMC Fam. Pract.* **7**:56.

91. Vernacchio, L., R. M. Vezina, A. A. Mitchell, S. M. Lesko, A. G. Plaut, and D. W. Acheson. 2006. Diarrhea in American infants and young children in the community setting: incidence, clinical presentation and microbiology. *Pediatr. Infect. Dis. J.* **25**:2–7.

92. Vinjé, J., and M. P. Koopmans. 2000. Simultaneous detection and genotyping of "Norwalk-like viruses" by oligonucleotide array in a reverse line blot hybridization format. *J. Clin. Microbiol.* **38**:2595–2601.

93. Wheeler, J. G., D. Sethi, J. M. Cowden, P. G. Wall, L. C. Rodrigues, D. S. Tompkins, M. J. Hudson, and P. J. Roderick for The Infectious Intestinal Disease Study Executive. 1999. Study of infectious intestinal disease in England: rates in the community, presenting to general practice, and reported to national surveillance. *BMJ* **318**:1046–1050.

94. Widdowson, M. A., A. Sulka, S. N. Bulens, R. S. Beard, S. S. Chaves, R. Hammond, E. D. Salehi, E. Swanson, J. Totaro, R. Woron, P. S. Mead, J. S. Bresee, S. S. Monroe, and R. I. Glass. 2005. Norovirus and food-borne disease, United States, 1991–2000. *Emerg. Infect. Dis.* **11:**95–102.

95. Wolfaardt, M., M. B. Taylor, H. F. Booysen, L. Engelbrecht, W. O. Grabow, and X. Jiang. 1997. Incidence of human calicivirus and rotavirus infection in patients with gastroenteritis in South Africa. *J. Med. Virol.* **51:**290–296.

96. Zintz, C., K. Bok, E. Parada, M. Barnes-Eley, T. Berke, M. A. Staat, P. Azimi, X. Jiang, and D. O. Matson. 2005. Prevalence and genetic characterization of caliciviruses among children hospitalized for acute gastroenteritis in the United States. *Infect. Genet. Evol.* **5:**281–290.

Food-Borne Viruses: Progress and Challenges
Edited by Marion P. G. Koopmans, Dean O. Cliver, and Albert Bosch
© 2008 ASM Press, Washington, DC

Emerging Food-Borne Viral Diseases

5

Erwin Duizer and Marion Koopmans

Severe acute respiratory syndrome (SARS) virus, avian influenza virus, and Nipah virus are examples of viruses for which food-borne transmission was not considered immediately but that illustrate the potential and the difficulty in assessing the potential for food-borne transmission. Is it possible to predict which viruses might be or become food-borne, based on known properties and comparison with members of the same family from the animal world? What are the factors involved in the emergence of food-borne viral disease? This chapter focuses on the basic information that has been gathered on these pathogens from the perspective of food safety.

EMERGING VIRAL INFECTIONS

In the second half of the 20th century, infectious diseases were considered to be a controllable problem (131). Epidemiological studies had found the mode of transmission for cholera, and drinking-water treatment was one of the first major contributors to improved health. The possibility of antibacterial therapy and the introduction of vaccines for preventing or mitigating infectious diseases further helped increase the health status of the population, although access to these health-promoting factors certainly is not uniformly available worldwide.

Since then, the situation has changed, particularly following the emergence of human immunodeficiency virus (45), a virus that was introduced into the population from an animal reservoir and subsequently spread across

ERWIN DUIZER AND MARION KOOPMANS, Laboratory for Infectious Diseases, Centre for Infectious Diseases Control, National Institute for Public Health and the Environment, 3720 BA Bilthoven, The Netherlands.

the world through sexual, blood-borne, and vertical transmission to an unprecedented scale (45). The extremely long incubation period allowed widespread dissemination of the virus before it was recognized that there was a problem. Even after it was recognized, it took a long time before this information was translated into widely accepted practical guidelines (35). This example clearly shows the difficulties we encounter when faced with a new disease. Similarly, after it was recognized that a prion disease of sheep had the potential to cause a fatal condition in cattle fed with meat and bone meal, it took 10 years before effective control measures were put in place in the European Union, highlighting the weaknesses of the system (100) (http://www.bseinquiry.gov.uk/report/volume16/chron2.htm). In this case, the total death toll in humans from the illness has remained fairly low, but the example highlights the tough balance between economic success and food safety (http://www.eurocjd.ed.ac.uk/vcjdworldeuro.htm).

Since then, several new infectious diseases have been discovered, with more or less media coverage. Table 1 lists viruses that were mentioned as emerging-disease threats in 1995 when the first issue of the *Emerging*

Table 1 Recent examples of emerging infections and factors contributing to their emergence and spread

Yr reported (reference)	Agent	Factors contributing to emergence	Factors contributing to spread
≤1995 (82)	Arenaviruses	Changing agriculture practices favoring rodents	Close contact, health care settings
	Bovine spongiform encephalopathy agent	Changing feed preparation	Complex food and feed chain
	Dengue virus (HF)	Urbanization, breakdown of vector control	Transportation of plants, vector, and infected people
	Ebola and Marburg viruses	Unknown	Local funeral practices
	Hantavirus	Increased contact with rodents, new virus?	
	HBV and HCV		Intravenous drug use, medical practices, mother-child transmission
	Human immunodeficiency virus	Bushmeat consumption, habitat encroachment	Sexual contact, intravenous drug use, medical practices, mother-child transmission
	Rift Valley fever virus	Dam building, agriculture, irrigation; possibly change of virus	
	Yellow fever virus	Conditions favoring mosquito spread	Transportation of vector

Table 1 (*continued*)

Yr reported (reference)	Agent	Factors contributing to emergence	Factors contributing to spread
1995 (26, 109)	Nipah and hendra viruses	Habitat encroachment leading to close human-animal contact	High-density pig production with low biosecurity
1996[a]	Highly pathogenic avian influenza virus HPAI H5N1	Viral evolution, urbanization, and large-scale poultry production; close animal-human contact	Legal and illegal trade in wild birds and meat
1996 (37)	G9P6 rotavirus	Viral evolution, close human-animal contact	Unknown
1997 (79)	HEV	Unknown	
2001 (44)	Circoviruses	Unknown	Unknown
2003 (90)	SARS virus	Habitat encroachment leading to close human-animal contact, local eating habits including bushmeat consumption	International travel, modern health care settings
2004 (59)	Avian influenza virus HPAI H7N7	Viral evolution, urbanization, and large-scale poultry production; close animal-human contact	International travel, modern health care settings
2004 (97)	Mimivirus	New detection methods	Unknown
2004, 2005 (125, 133)	Other coronaviruses	New detection methods	Contact, endemic viruses
2005 (3)	Bocavirus	New detection methods	Endemic viruses

[a]http://www.WHO.INT/CSR/disease/avian_influenza/Timeline 2008_0_02.pdf

Infectious Diseases journal was launched. Since then, several new viruses have been added to the list. In this chapter, viruses are considered emerging pathogens if their incidence has increased within the past two decades or threatens to increase in the near future.

FACTORS CONTRIBUTING TO THE EMERGENCE OF NEW DISEASES

Although it is difficult to prove, the consensus is that the potential for emergence of novel infectious diseases has been increasing in the past decades because of multiple factors, mostly directly or indirectly related to human behavior. Factors can broadly be grouped in the following categories: human population growth and migration, globalization of the market for food, ecological factors, and pathogen-related factors.

Human Population Growth and Migration

Human actions are the most important factors driving changes in infectious-disease risks. The world has experienced an explosive population growth over the past decade, resulting in a rapid increase in the demand for (cheap) animal protein such as poultry meat (FAOSTAT; http://faostat.fao.org). This increase is particularly seen in Asia, with China as a leading country. The meat consumption per person in the developing countries went from 11 to 23 kg between approximately 1975 and 1995 (http://www.fao.org/), and this is coupled to the increasing number of megacities developing, particularly in Asia, due to population growth (Fig. 1). The resulting increased production of animal protein is not just a numerical change: animal husbandry practices differ greatly from one geographical region or country to another, with a large proportion of animals still being raised in small-scale production systems (http://www.fao.org/). In these holdings, compliance with basic biosecurity measures (such as preventing the introduction of animals of different origin into a flock) is low; therefore, diseases may emerge rapidly, especially if live-animal trade is involved as well (18).

Planned and especially unplanned migration leads to situations in which populations gather, often with suboptimal hygienic conditions, in regions that have difficulty in coping with the increasing demand on the local infrastructure such as the food and water supply. These circumstances may result in introductions of viruses that under healthier living conditions do not pose much of a problem. A recent example is the large hepatitis E outbreak with over 2,600 cases in a camp with 78,000 refugees in Darfur, Sudan (10). Hepatitis E virus (HEV) has been known as a fecal-oral pathogen endemic in many developing countries, with occasional infections that typically are relatively mild and self limiting. Studies of HEV in high-income countries

Figure 1 Number of cities with a population of more than 1 million in 1900 and more than 5 million in 2000, showing a demographic change underlying the increases in the risk of emergence of infectious diseases. SA, South America. Source: U.N. Food and Agriculture Organization (http://www.fao.org).

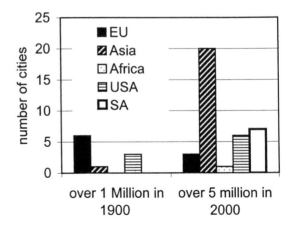

with hygienic living conditions suggest that it does not cause much clinical disease or large outbreaks in these countries. In contrast, large outbreaks have been described following failures of water treatment in developing countries. In these instances, the impact may be high, with significant mortality (10, 40). Interestingly, HEV genotype 3 infections in high-income countries seem to be restricted largely to persons with underlying disease. Finally, travel facilitates the spread of infectious diseases. The actual routes a disease will take are determined by how individuals travel within and between regions; current travel patterns enable contagious diseases to spread to the far corners of the globe at alarming rates. The SARS epidemic made this very clear (16).

Globalization of the Market for Food
The global trade of foods has more than tripled in the past two decades, with big differences by region (ranging from 1.7-fold for the United States to >6-fold for China) (Fig. 2). While regulations are in place to monitor the microbiological quality of food, the criteria that are used have been developed for monitoring bacterial contamination and do not accurately reflect the presence or absence of viruses (103). The early detection of potential diffuse outbreaks due to contaminated food therefore relies on surveillance and international exchange of the findings from the surveillance (58).

Figure 2 Exports of foods from different countries around the world, 1980 to 2006. Source: World Trade Organization (www.wto.org).

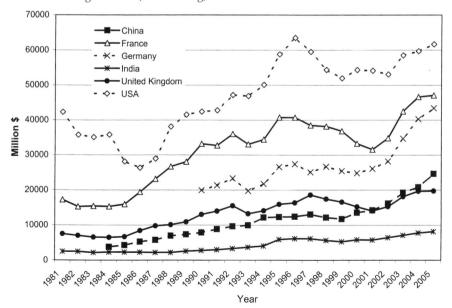

The surveillance of food-borne viral disease has not received high priority in countries across the world, and this has been declared an important data gap by a World Health Organization expert consultation (134). In addition, traceback of contaminated commodities may be problematic (86). Risks across the world may vary considerably, depending on the nature and origin of foods consumed and food production practices (106). An extreme example is the deluxe coffee made from berry seeds that have been roasted and ground after being passed through the intestinal tract of luwaks (an Indonesian civet cat) or weasels, practices that are found in Southeast Asia (http://en.wikipedia.org/wiki/ Kopi_Luwak).

Ecological Factors
Many of the recent examples of emerging infections are zoonoses, and the consensus is that animal reservoirs are a major contributor to the emergence of new infectious diseases. Taylor et al. (118) made a detailed catalogue of currently known pathogens, including those causing emerging infections, and concluded that 75% of these infections were zoonotic. According to this overview, protozoa and viruses were most likely to emerge, irrespective of the mode of transmission. The demographic changes resulting from population growth lead to rapid deforestation and encroachment of humans into wilderness areas, with resulting risk of contact with infections commonly found in wild animals (24). Nipah virus was thought to be introduced into Malaysian pigs by fruitbats that migrated from Indonesia when their forest habitats were substantially reduced by deforestation for pulpwood and industrial plantation (27). Wildlife trade and translocation, live-animal and bushmeat markets, consumption of exotic foods, development of ecotourism, and ownership of exotic pets contribute to this risk as well (29, 98).

Pathogen-Related Factors
In addition to the factors listed above, pathogen-related factors may influence the risk of emergence of new infectious diseases. Viruses, especially RNA viruses, are particularly flexible because the replication process introduces errors with each round of replication, so that the progeny consists of a swarm of mutant viruses. Application of selective pressure (e.g., inoculation of a new host) may lead to the selection of a mutant that is favored under these new conditions. In addition to this, viruses may change by exchange of genetic material during replication (through recombination or reassortment), resulting in the generation of new viruses (see chapter 6). Rotavirus is an example of such a mechanism in a food-borne virus; new viruses occasionally emerge in the human population when new surface proteins are acquired through reassortment of genes of human rotaviruses with viruses from

another source, thought to be zoonotic (63, 96). The food-borne mode of transmission may actually favor the generation of recombinant viruses when the food is contaminated with sewage. This occurs quite often for products like molluscan shellfish (13, 66). Another risk category is fresh produce, e.g., berries or lettuce, irrigated with contaminated water while growing (15).

POTENTIAL FOR EMERGENCE AND SPREAD OF NEW VIRAL DISEASES VIA THE FOOD CHAIN

Viruses have long been somewhat neglected in the food safety debate. Currently used control measures focus on bacterial contamination, which does not rule out or detect viral presence. The examples for norovirus discussed elsewhere in this book illustrate that food-borne virus transmission is quite common, with outbreaks occurring on a regular basis (58). The largest category here is the food handler-associated outbreak, in which viruses shed by an infected individual are passed into food, which then becomes a vehicle (33, 46). Viruses that are stable in the food environment may then be transmitted to other persons following ingestion of contaminated food (33, 84, 89). At present, there is little control of food-borne viral transmission, leading to the conclusion that, if new viruses emerged and spread via food, this would not be immediately recognized or stopped by existing control measures. Therefore, it seems prudent to prepare for the potential of such an event.

Prerequisites for Food-Borne Transmission

To be able to discuss the prerequisites for food-borne transmission, we must define food-borne transmission: it is defined as the transmission of a pathogen by consumption of a pathogen-contaminated food item. What we call the main food-borne viruses, i.e., hepatitis A virus (HAV) and noroviruses, are viruses that are specific for humans and therefore multiply in and infect only humans. This means that these infectious pathogens are normally spread by person-to-person contact. However, since the viruses can be shed in large numbers, are stable in the environment, and survive many food production processes, infected food handlers can contaminate food, which then serves as a vehicle for onward transmission. This route is described here as food handler transmission.

Another route that exists for some pathogens involves infection of animals in the food chain, which then leads to human exposure. This route is termed zoonotic food-borne transmission. Direct zoonotic food-borne transmission occurs if edible parts are contaminated by a zoonotic pathogen that causes infection when food preparation does not involve a cooking step that results in inactivation. Since viruses tend to be quite host specific, this route is rare for viruses, but if it occurs it could be one way of introducing new pathogens into humans. Zoonotic food-borne transmission can also be indirect if infected animals shed virus in their excreta and thus contaminate other food products, as exemplified for Nipah viruses. When emerging infections are discussed, food-borne zoonotic transmission is relevant, since 75% of emerging infectious agents are thought to be of zoonotic origin (118). Finally, food contamination via environmental spread (sewage, sewage-contaminated irrigation water) is a possible route to introduction of viruses into the food chain.

It is clear that zoonotic food-borne transmission and food handler transmission are different routes, involving different sets of characteristics of the viruses for successful employment. Zoonotic food-borne transmission is well known for several bacteria; this route is, however, rare for viruses. HEV is an example of a virus that causes infection via direct zoonotic food-borne transmission (120). It infects animals such as pig, wild boar, and deer; consumptions of infected liver can lead to infection and disease in humans. This route has also been suggested to have occurred after consumption of raw duck blood during the H5N1 avian influenza epizootic in Vietnam in 2005 (http://www.cdc.gov/travel/other/avian_flu_ig_americans_abroad_032405.htm). This chapter looks more closely at virus properties contributing to the likelihood of food handler transmission and zoonotic food-borne transmission.

Features That Favor Food Handler Transmission

Food handler transmission is known for HAV and norovirus but also for other viruses such as adenoviruses, aichiviruses, astroviruses, enteroviruses, rotaviruses, and sapoviruses. This transmission route first requires that viruses be shed by humans and be infectious for humans by the oral route. Additionally, the viruses must be able to survive eventual postcontamination food production processes. This means that they are stable in the environment and in food matrices unless contamination occurs just before consumption (i.e., close to the "fork"). Furthermore, since food handler contamination of foods most probably results in a low level of contamination, a low minimal infectious dose is an advantage for a virus. Most of the viruses listed above are small spheres, with a single-stranded positive RNA genome and without envelope (Table 2). Only the adenoviruses are DNA viruses, and only the coronaviruses possess an envelope. Clearly, the lack of an envelope has a great influence on the stability and resistance (52) of viruses and seems to be a factor contributing to the likely success of viruses as food handler-transmitted pathogens.

Table 2 Overview of taxonomic groupings of mammalian viruses, their zoonotic potential, and an assessment of their theoretical (or reported) potential for food-borne transmission based on virus properties

			Theoretical potential for:		
Taxonomic group	Family	Documentation of zoonotic transmission	Food-borne transfer[a]	Zoonotic food-borne transmission[b]	Food handler transmission
dsDNA[g]	Adenoviridae	No	Possible	Unlikely	Possible
	Asfarviridae	No	Unlikely	Unlikely	Unlikely
	Hepadnaviridae	Yes	Unlikely	Unlikely	Unlikely
	Herpesviridae	Yes	Unlikely	Unlikely	Unlikely
	Papillomaviridae	No	Possible	Unlikely	Possible
	Polyomaviridae	No	Possible	Unlikely	Possible
	Poxviridae	Yes	Unlikely	Unlikely	Unlikely
ssDNA[g]	Circoviridae	No	Possible	Unlikely	Possible
	Parvoviridae	No	Possible	Unlikely	Possible
dsRNA	Birnaviridae	No	Possible	Suspected	Possible
	Reoviridae	Yes	Proven[c]	Proven[c]	Suspected
− strand RNA	Arenaviridae	Yes	Unlikely	Suspected	Unlikely
	Bornaviridae	No	Possible	Unlikely	Unlikely
	Bunyaviridae	Yes	Unlikely	Unlikely	Unlikely
	Filoviridae	Yes	Unlikely	Unlikely	Unlikely
	Orthomyxoviridae	Yes	Possible	Suspected	Possible
	Paramyxoviridae	Yes	Possible	Proven[d]	Possible
	Rhabdoviridae	yes	Unlikely	Unlikely	Unlikely
+ strand RNA	Arteriviridae	No	Proven[e]	Unlikely	Unlikely
	Astroviridae	No	Proven[c]	Unlikely	Proven[c]
	Caliciviridae	Yes	Proven[c]	Possible	Proven[f]
	Coronaviridae	Yes	Proven[c]	Possible	Possible
	Flaviviridae	Yes	Possible	Unlikely	Unlikely
	Hepeviridae	Yes	Proven[c]	Proven[c]	Possible
	Picornaviridae	Yes	Proven[c]	Proven[c]	Proven[c]
	Retroviridae	Yes	Unlikely	Unlikely	Unlikely
	Togaviridae	No	Unlikely	Unlikely	Unlikely

[a]Transfer indicates the physical transfer of the virus to the next host. Unlikely indicates that the virus is unlikely to be present on food or unlikely to remain infectious until consumption. Possible indicates that based on the virus features, this route cannot be ruled out. Proven indicates that published reports on this route exist.

[b]Transmission includes the possibility to infect, thus including infectivity and host specificity.

[c]Reviewed in reference 38.

[d]Reference 70.

[e]Reference 73.

[f]Reviewed in reference 33.

[g]ds, double stranded; ss, single stranded.

Factors Favoring Zoonotic Food-Borne Transmission

So far the number of recognized zoonotic food-borne viral infections is very limited, and to date it is not clear if this is due to huge underreporting or to the fact that these infections actually do happen only rarely. Recently, direct zoonotic food-borne transmission was reported for HEV that had caused infections after consumption of contaminated undercooked livers of wild boar and deer (68, 119, 120). Since then, several groups have started to screen for the presence of HEV in meat sold for human consumption; as a result, viral RNA has been found in pig livers in Japan (135), the United States (36), and The Netherlands (12a). HEV is discussed in depth elsewhere in this volume. Other viruses for which food-borne zoonotic transmission has been suggested are SARS coronavirus (106) and avian influenza virus (H5N1) (http://www.who.int/foodsafety/micro/avian/en/index.html).

Factors such as having a wide host range and being able to replicate in the human gastrointestinal tract, as well as being stable, determine the probability that viruses may be transmissible via the zoonotic food-borne route. In general, heating at temperatures above 60°C will suffice to inactivate viruses present in meat, but the efficacy of heat treatment depends on the composition of the food matrix and may be poor (9, 57). In addition, meat products such as chorizo and Italian salami are minced and then treated for preservation by salting, smoking, and/or drying. The efficacy of these processes for virus inactivation has not been systematically studied. Survival of foot-and-mouth disease virus (a nonenveloped RNA virus belonging to the *Picornaviridae*) and hog cholera virus (an enveloped RNA virus belonging to the *Flaviviridae*) in Italian salami for up to 75 days after manufacture has been demonstrated. Clearly, the survival of the hog cholera virus was longer than the useful life of many kinds of salami. Foot-and-mouth disease virus was apparently greatly affected by the drop in pH that occurs in meat after slaughter (87, 88).

VIRUSES TO BE CONSIDERED FOR FOOD-BORNE TRANSMISSION

If we assume a minimal set of viral features required for efficient food-borne transmission, it would include (i) small (spherical) size, (ii) lack of envelope, and (iii) infectivity for humans or at least mammals. Applying these criteria to all known virus families (http://www.ncbi.nlm.nih.gov/ICTVdb/Ictv/index.htm), we find that *Adenoviridae, Astroviridae, Birnaviridae, Caliciviridae, Circoviridae, Hepeviridae, Papillomaviridae, Parvoviridae, Picornaviridae, Polyomaviridae,* and *Reoviridae* are the virus families containing members likely to survive outside the host. Representatives of the families *Adenoviridae*, *Astroviridae*, *Caliciviridae*, *Hepeviridae*, *Picornaviridae*, and *Reoviridae* have clearly been documented to be food-borne viruses and are not discussed further in this chapter (they were recently reviewed in reference 38). Instead, we focus on the other virus families: *Birnaviridae*, *Circoviridae*, *Papillomaviridae*, *Parvoviridae*, and *Polyomaviridae*.

Birnaviridae

The *Birnaviridae* are small, double-stranded RNA viruses with a segmented genome and without an envelope. The family consists of four genera of which only one, *Picobirnavirus*, is known to infect mammals and humans. Picobirnaviruses have been detected in the stools of several birds, mammals, and human gastroenteritis patients (71, 75). Even though the role of pico-birnaviruses as a cause of gastroenteritis has not been established unambiguously, the viruses are now considered opportunistic pathogens causing diarrhea in immunocompromised patients such as human immunodeficiency virus-infected persons and organ transplant recipients (7). Currently there is a clear need for information on the transmission routes and zoonotic potential of these viruses; however, fecal-oral transmission is likely to be relevant and food-borne transmission cannot be excluded.

Circoviridae

Circoviridae are small viruses with a single-stranded circular DNA genome of less than 4 kb and without an envelope. This family consists of three genera, *Anellovirus*, *Circovirus*, and *Gyrovirus*. The human circoviruses belong to the genus *Anellovirus* and can be subdivided into torque teno virus (TTV) and the torque teno mini virus (TTMV) (47). TTV was first detected in a patient with hepatitis, but currently no disease or pathology is related to TTV infection. The distribution of the virus is presumed to be global, and the prevalence of TTV DNA in human sera is very high (8). TTV is also named transfusion-transmitted virus, but other modes of transmission, among them fecal-oral, are suspected. TTV and TTMV sequences have been detected in nonhuman primates but also in chickens, pigs, cows, sheep, and camels (4, 47). Moreover, a common source for TTV in camels and humans was suggested to exist in the United Arab Emirates (4).

The porcine circovirus (genus *Circovirus*) could be detected in fecal swabs (108), and it is suggested that it can be transmitted by the oral route through milk (111). Data on the food-borne transmission potency of human circoviruses are nonexistent, but apparently members of the *Circoviridae* family are capable of exploiting this route.

Papillomaviridae

Papillomaviridae is a family of double-stranded DNA viruses without an envelope. These viruses normally infect epithelial cells of animals and humans. Transmission can occur through direct contact (via the skin or sexually) or indirectly via contaminated fomites. Papillomaviruses are highly persistent in the environment and can remain infectious on contaminated objects for over a week (102). Following infection, cells may become immortalized to form hyperproliferative lesions known as warts, papillomas, or condylomas. Typically, papillomavirus-induced lesions are benign and self-limiting, and they spontaneously regress. However, some papillomavirus types are linked to malignant tumors. The best known of these are the human papillomavirus types 16 and 18, which cause cervical carcinoma.

Different papillomaviruses have been identified in (food) animals such as elk, cattle, and sheep and are grouped as a separate genus within the family *Papillomaviridae*. In cattle, papillomaviruses can also infect fibroblasts and induce benign fibroepithelial tumors. Since these tumors are often located on the teats, it is likely that viruses could be transmitted to milk (72). Experimentally, bovine papillomaviruses associated with a form of cystitis have been transmitted to recipient animals when blood from animals passing bloody urine was transferred (115). Cross-species transmission has been documented for bovine papillomaviruses that may cause sarcoids (benign fibroblastic skin tumours) in horses (20). In cattle, papillomas can be found in the upper gastrointestinal tract following oral infection (123). Malignant tumors are thought to arise occasionally from these infections (17). Bovine papillomavirus was also shown to be quite resistant to heat, with less than 3 $\log_{10}$ inactivation at 74°C for 1 h (102). There is no evidence for zoonotic transmission of these viruses.

Parvoviridae

Parvoviridae is a family of small, single-stranded DNA viruses without an envelope. The best-known human parvovirus is parvovirus B19, which causes erythema infectiosum (fifth disease, slapped-cheek syndrome). This virus is transmitted mainly via respiratory secretions, but it can also be transmitted via blood or by the maternal-fetal pathway (31). The respiratory excretions can lead to environmental, fomite, and food contamination and subsequently to infection of other persons. Parvoviruses are also thought to be the causative agents of gastroenteritis due to shellfish consumption in a large outbreak in the United Kingdom in 1977 (38).

Other recently discovered parvoviruses are from the genus *Bocavirus* (3). Human bocavirus was thought to cause only a respiratory infection, but recent reports show that gastroenteritis and fecal shedding are common during bocavirus infection in children (81, 128).

Animal parvoviruses are often related to diarrhea. Parvoviruses have been isolated from stools of dogs, cats, pigs, rabbits, horses, and mink; fecal-oral transmission is thought to be a common route of transmission (2, 113). Thus far, no zoonotic transmission of parvovirus has been reported.

Polyomaviridae

The human pathogens JC and BK virus belong to the *Polyomaviridae*, which is a family of nonenveloped viruses with a DNA genome. The viruses are widely distributed throughout the world, although the pattern of JCV infection appears to vary among populations (56). People acquire infections typically at a young age, and up to 90% of the adult population has antibodies to the virus. No acute disease has been associated with infection in immunocompetent hosts, but the viruses remain latent after primary infection. Illnesses associated with these viruses develop when people become immunocompromised, particularly when cellular immunity is decreased, either from iatrogenic causes (in transplant recipients) or following human immunodeficiency virus infection. The human polyomaviruses BKV and JCV cause, respectively, hemorrhagic cystitis in recipients of bone marrow transplants and progressive multifocal leukoencephalopathy in immunocompromised patients (1). The mode of transmission of these viruses is unknown, but virus is shed in the urine, suggesting that this body fluid plays a role in transmission (6). A related virus from rhesus macaques, simian virus 40 (SV40), has been shown to cause tumors in rodents, sparking discussions about the possible role of this and other polyomaviruses as a cause of human cancers. SV40 was inadvertently inoculated into an estimated 10 million to 30 million Americans as contaminant of a poliovirus vaccine produced between 1954 and 1962. Many epidemiological and molecular pathogenesis studies have been conducted to identify potential cancer risks since this "natural" experiment began, but so far these studies have remained inconclusive, although SV40 has been detected in a wide range of tumors (116).

Since polyomaviruses are widespread, are shed in urine, can be found in environmental samples, and—based on their structure—are likely to be stable in the environment, their presence as food and water contaminants cannot be ruled out (11, 12). However, given their omnipresence, this cannot be considered an emerging disease risk. It is possible that other polyomaviruses exist, for example in food animals, but the narrow host range makes crossover of such viruses into humans unlikely. The potential association between JCV and cancer is a cause for concern and, if proven, would point to an immediate need to identify the possible modes of transmission, including the risk for food-borne transmission (132).

OTHER VIRUS FAMILIES WITH REPORTED FOOD-BORNE TRANSMISSION

The virus families above were selected based on viral characteristics that would make them candidates for food-borne transmission. There are also viruses that are known or thought to be food borne but lack the general features of food-borne viruses (small spheres, no envelope, infectious to humans). Among these are viruses from the families *Coronaviridae*, *Flaviviridae*, *Orthomyxoviridae*, and *Paramyxoviridae*.

Coronaviridae

Since the emergence of the SARS coronavirus (CoV) in 2002, there has been a lot of debate on the role of the fecal-oral transmission route in the spread of this virus. Soon after the identification of a CoV came the recognition of gastrointestinal symptoms and the detection of SARS CoV in the feces of many patients (23, 67). Viruses of the family *Coronaviridae* are enveloped viruses with a positive-strand RNA genome. The viruses can measure up to 170 nm in diameter, and their 27- to 32-kb genomes not only are the largest of all RNA viruses but also display a very high recombination frequency (62, 107). CoV strains cause a wide variety of diseases in different animal species, and interspecies and zoonotic transmission of CoV has been reported (43, 101). Until the emergence of SARS CoV, human CoV strains were usually associated with relatively mild respiratory infections ("the common cold") and only occasionally with enteric disease (49). SARS was characterized initially (early 2003) as an atypical pneumonia that presented with severe respiratory symptoms and a high fever. Since then, additional symptoms such as general malaise, muscle stiffness, headaches, rash, and diarrhea have been reported (22, 67). Among the first patients were many persons working at one of the Chinese live-animal markets and health care workers. Currently, bats are thought to be the natural reservoir for SARS CoV, but other animals such as civets and raccoons are likely to have played a role as host in the emergence of the SARS epidemic (110, 130).

Even though SARS CoV is a virus with an envelope and its main transmission route is respiratory, food-borne transmission cannot be excluded. First, the gastrointestinal tract is clearly involved in a large fraction (about 40%) of persons infected with SARS CoV: replication in the small and large intestine, diarrhea, and fecal shedding do occur (67). Moreover, the outbreak in Amoy Garden in Hong Kong is thought to have been caused by fecal shedding that led to airborne and environmental transmission (30, 78, 95). Interestingly, replication of SARS CoV in the nasopharynx was related to diarrhea (22). This might suggest that the gastrointestinal tract is infected by viruses swallowed after replication in the nasopharynx. This would then imply that SARS CoV strains are able to survive the acidic milieu in the stomach, but experimental data on SARS CoV stability at low pH are not available yet.

Even though CoV does have an envelope, it is relatively stable in the environment and can survive for days on fomites and surfaces (SARS CoV [32]) or in suspension (mouse hepatitis virus, [104], HCoV-229e [64]). In a comparative study, SARS CoV was found to be more stable in suspension than in a dried state; however, in both cases it was more stable than HCoV-229E (95).

For SARS CoV, two food-related transmission routes should be considered. The first is the direct zoonotic food-borne transmission by consumption of any of the nonhuman hosts of the virus. Since SARS CoV infectivity is rapidly reduced at higher temperatures (56°C and higher) (http://www .who.int/foodsafety/publications/micro/en/sars_madrid.pdf), infection by this route depends on consumption of raw contaminated products originating from bats, raccoons, or civets. Even though raw or undercooked products of these animals are said to be consumed, no proof of infection by this route exists.

The other route is food handler transmission. Obviously, foods can be contaminated by respiratory (coughing, sneezing, breathing) and fecal and urinary shedding. For this route it is important to know if shedding occurs after recovery from illness and if asymptomatic shedding occurs. Nasopharyngeal shedding of SARS CoV is reported for almost 60% of patients, and the viral load of SARS CoV in nasopharyngeal specimens may vary from undetectable to 10^8 RNA copies per ml. Shedding in sputum can last an average of 21 days, with subsets of patients shedding virus in sputum for over 6 weeks (21, 69). Additionally, SARS CoV was detected in nasopharyngeal swabs from subclinically infected health care workers (48). Shedding of SARS CoV in stools (an average of 27 days) is reportedly longer than in sputum (an average of 21 days), with a subset of patients having detectable SARS CoV in stools for over 10 weeks (69). Urinary secretion of SARS CoV seems to occur in a smaller fraction of patients (5%) but may last at least as long as fecal shedding (65). Clearly, the long periods of shedding point at a window of opportunity for the viruses to be transmitted via food handlers. When the emergence of a CoV is reported or noticed, food-borne transmission should be considered, both in studies to determine the emergence and spread of the virus and in putting countermeasures into place.

Flaviviridae

Flaviviridae is a family of positive sense RNA viruses, with virions ranging in size from 40 to 60 nm with an envelope. The family consists of three genera: the genus *Flavivirus* to which the yellow fever virus, tick-borne encephalitis virus (TBEV), dengue virus, Japanese encephalitis virus, and West Nile virus (WNV) belong; the genus *Pestivirus*, to which the bovine viral diarrhea and classical swine fever viruses belong; and the genus *Hepacivirus*, to which the hepatitis C viruses (HCV) belong. Only viruses belonging to the *Hepacivirus* and *Flavivirus* genera are known to infect humans. The hepacivirus HCV is mainly blood borne (76, 93). No zoonotic transmission or examples of food-borne transmission are known.

Viruses belonging to the genus *Flavivirus* are mainly arthropod-borne viruses, but examples of zoonotic food-borne transmission have been reported. TBEV is transmitted by ticks (*Ixodes* sp.) from its natural hosts, mostly rodents, to humans or, for example, to cows, sheep, and goats. In these animals the viruses can be shed via milk, and consumption of contaminated raw milk can lead to infection and a disease called "biphasic milk fever" in humans (39, 54, 61, 127). Moreover, infectious TBEV is found in yoghurt, butter, and cheese; the virus is able to survive in gastric juice for 2 h. The high resistance to acid is not concomitant with a generalized high resistance to inactivation. Due to the lipid envelope, TBEV is readily inactivated by heat treatment, detergents, and organic solvents (14). Even though a viremic phase is common during a TBEV infection in several animal species, food-borne infection via contaminated meat or organs is unlikely because of the fast virus inactivation at elevated temperatures. TBEV can cause a variety of clinical symptoms after an incubation period of 7 to 14 days. Common early symptoms are fatigue, headaches, and pain in the neck, back, and shoulders. These may progress to a sudden onset of the classic symptoms such as fever; nausea and vomiting; severe muscle pain in the neck, back, shoulders, and limbs; and encephalitis. The case fatality rate in Europe is in general low (0.5 to 1.5%) but may differ depending on the virus strain and/or geographic region (39, 41, 42).

WNV is an arthropod-borne flavivirus that normally infects birds and uses mosquitoes as its vector. It can, however, infect many other animals such as mice, hamsters, and horses, as well as humans. WNV was already recognized in the 1930s, but infections were mostly asymptomatic or mild. In the early 1990s, the frequency and severity of infection in humans increased, and the virus emerged in the United States in 1999. Human infections can be asymptomatic, but they can also cause fever and neurological symptoms (meningoencephalitis); severe disease may be fatal. Mammals are considered dead-end hosts for WNV (92), but transmission of WNV from human to human via blood transfusion has been reported (91). Even though vector-borne transmission is the main route for WNV, food-borne infection has been reported to occur in alligators, hamsters, and cats after consumption of infected mice (5, 55, 105) and ingestion of infected birds or insects has resulted in infections of birds (126). Additionally, alligators have been infected after consumption of parts of an infected horse (51). Infection of a human baby by breast-feeding is suggested to have occurred in the United States (19). Furthermore, high titers of infectious virus have been detected in cloacal swabs and feces of passeriforms (perching birds) and in urine of small rodents (122, 126). In summary, ingestion of WNV-contaminated products and foodstuffs is not an unlikely event and in theory could lead to infection. Whether this mode of transmission plays a role in the epidemiology of WNV remains to be seen.

Orthomyxoviridae

The current epizootic of highly pathogenic avian influenza (AI) is caused by a strain (H5N1) of the genus influenza A viruses. Influenza viruses belong to the *Orthomyxoviridae* family, consisting of large (~300-nm), minus-strand RNA viruses with a segmented genome and an envelope. Influenza viruses are found in a wide range of vertebrate hosts. The natural hosts of avian influenza viruses are waterfowl, but the viruses are found in a wide range of other birds (85). Additionally, AI has spread to several kinds of mammals, including humans, pigs, and cats (60, 74, 112, 121).

Even though the physical characteristics of these viruses do not favor efficient food-borne transmission, the large scale of the A/H5N1 epizootic in widely consumed birds such as chicken, turkeys, ducks, and geese, and consequently the huge impact if only a small chance of food-borne transmission existed, warrants a closer look a the possibility of food-borne transmission of AI. The risk or possibility of food-borne transmission has been studied by such organizations as INFOSAN (International Food Safety Authorities Network), EFSA (European Food Safety Authority) and the International Association for Food Protection (http://www.who.int/foodsafety/micro/avian/ en/index .html; http://www.efsa.europa.eu/en/science/biohaz/biohaz_documents/ 1412.html; http://www.foodprotection.org/AvianFluPresentations/Avian% 20Influenza%20Scientific%20Brief.pdf).

Several features of AI viruses contribute to the likelihood of food-borne transmission. First, the viruses are able to cross the species barrier from birds to humans and from birds to pigs to humans. Second, AI viruses have been found in several edible parts of infected birds. Meat, organs, blood, and eggs all can contain infectious AI viruses (94, 117, 134). Eggs can be contaminated on the outside and on the inside, in both egg whites and yolk.

Therefore, infectious AI viruses may be ingested by humans if foods are not cooked properly (to a temperature above 70°C in all parts of the food item). Additionally, infections in tigers and leopards have occurred following consumption of raw chicken, and cats have also contracted AI when experimentally fed infected hatchlings (53, 60, 114). Virus was detected in the intestinal tract, brain, lungs, liver, kidneys, and spleen of cats. It is, however, not yet clear if viruses are able to infect humans if ingestion is the only route. Therefore, information on AI virus stability and the presence of AI virus receptors in the human gastrointestinal tract is essential. Currently, it is thought that the nonoptimal receptor-binding specificity of AI viruses hampers their replication in humans (77). However, it is clear that humans can be infected by H5N1 viruses; interestingly, replication in humans occurs exclusively in the lungs and the intestine (124), while replication in birds and, for example, cats is disseminated (60).

Paramyxoviridae

Nipah viruses, together with the hendra viruses, constitute the genus *Henipavirus* of the family *Paramyxoviridae*. The Nipah viruses are relatively large (120- to 500-nm), single-stranded, nonsegmented RNA viruses with an envelope. Symptomatic infections are characterized by influenza-like illness (high fever and muscle pain) that may progress to febrile encephalitis, convulsions, coma, and even death in 40 to 75% of patients with symptomatic infections (26, 34). Asymptomatic infections do occur. Nipah viruses have been detected in respiratory secretions and urine of patients and of infected pigs (28). The relevance of person-to-person transmission seems to differ between the outbreaks documented in Singapore and Malaysia (no indication of person-to-person transmission) and the outbreaks that occurred in Bangladesh. In the two outbreaks in Bangladesh, person-to-person transmission was suspected even though none of the health care workers involved showed any sign of Nipah infection (50). Contact with infected pigs or with an ill cow was found to be an important risk factor for human infection (26, 50). Direct contact with fruit bats (*Pteropus* sp.) or their excretions (urine and saliva) is another risk factor. Currently, fruit bats are considered the natural host for Nipah viruses, while several other animals, such as pigs and cats, can be efficiently infected too (80). The unusually broad host range of Nipah viruses facilitates zoonotic transmission (83).

Application of food safety criteria for Nipah viruses suggests that, based on their physical characteristics, these viruses are not candidates for efficient food-borne transmission. However, indirect zoonotic food-borne transmission by means of fruits contaminated by bat saliva or urine was considered to be a major source of infection for pigs (25), and an outbreak of Nipah virus encephalitis in humans was suggested to have resulted from Nipah virus-contaminated fresh date sap (70). It is also reported that Nipah viruses can survive for days in fruit juice and bat urine (70).

Additionally, local food consumption practices may contribute to disease emergence: in parts of Asia, the meat of several bat species, including the natural host of Nipah virus, *Pteropus* sp., is consumed. Bat blood is considered an aphrodisiac, and consumption of untreated bat blood is known to occur (99, 129). Since the RNA of Nipah viruses is detected in blood, consumption of improperly cooked bat products such as meat and blood may be a risk factor for infection.

APPROACH TO INCREASE THE LIKELIHOOD OF DETECTING FOOD-BORNE TRANSMISSION OF VIRUSES WITH CURRENTLY OPERATING SURVEILLANCE SYSTEMS

The question now is what can be done to improve the likelihood of detecting food-borne transmission. Virus detection in foods in still troublesome; only very limited resources are available, and it is unlikely that routine monitoring for viruses in foods will either be available shortly or be effective (33). Furthermore, there is the problem that food-borne transmission of viruses has only recently become appreciated and then only for HAV, norovirus, and the other viruses commonly found in filter-feeding shellfish. Therefore, the most important message is that the possibility of food-borne transmission should be seriously considered for every emerging pathogen, regardless of the clinical presentation. Food-borne transmission of viruses is often masked by other transmission routes that may be used and may be more important, such as person-to-person (HAV and norovirus) or respiratory (influenza virus, SARS CoV) transmission.

Based on viral characteristics, viruses belonging to families containing known food-borne pathogens and viruses structurally related to these families are highest on the list (*Adenoviridae, Astroviridae, Birnaviridae, Caliciviridae, Circoviridae, Hepeviridae, Papillomaviridae, Parvoviridae, Picornaviridae, Polyomaviridae*, and *Reoviridae*). In addition, however, this chapter has presented a number of examples of viruses that would be considered unlikely candidates for food-borne transmission based on their physical characteristics but still were reported to be food borne, such as AI and Nipah viruses.

CONCLUDING REMARKS

In summary, when considering viral properties such as genome type, host range, and presence of envelope, as well as a review of literature on food-borne transmission for animal viruses within the same families, the list of virus families and species that are more likely to be associated with food-borne transmission resulting in infection of humans can be narrowed down to 11 families. These include viruses for which the potential for food-borne transmission has never been documented or even studied, such as polyomaviruses with their unresolved mode of transmission. However, occasional food-borne transmission has been documented for viruses belonging to other families as well, although this is likely to be limited to problems following end-of-chain contamination. Nevertheless, the potential for unexpected introduction via the food chain should be considered with every new virus problem. Among the viruses discussed in this review, *Flavivirus*, *Orthomyxovirus*, and *Coronavirus* have the highest potential impact, and the possibility of food-borne transmission should be considered in preparedness planning for these pathogens.

REFERENCES

1. **Ahsan, N., and K. V. Shah.** 2006. Polyomaviruses and human diseases. *Adv. Exp. Med. Biol.* **577:**1–18.
2. **Allan, G. M., and J. A. Ellis.** 2000. Porcine circoviruses: a review. *J. Vet. Diagn. Investig.* **12:**3–14.
3. **Allander, T., M. T. Tammi, M. Eriksson, A. Bjerkner, A. Tiveljung-Lindell, and B. Andersson.** 2005. Cloning of a human parvovirus by molecular screening of respiratory tract samples. *Proc. Natl. Acad. Sci. USA* **102:**12891–12896.
4. **Al-Moslih, M. I., H. Perkins, and Y. W. Hu.** 2007. Genetic relationship of Torque Teno virus (TTV) between humans and camels in United Arab Emirates (UAE). *J. Med. Virol.* **79:**188–191.
5. **Austgen, L. E., R. A. Bowen, M. L. Bunning, B. S. Davis, C. J. Mitchell, and G. J. Chang.** 2004. Experimental infection of cats and dogs with West Nile virus. *Emerg. Infect. Dis.* **10:**82–86.
6. **Berger, J. R., C. S. Miller, Y. Mootoor, S. A. Avdiushko, R. J. Kryscio, and H. Zhu.** 2006. JC virus detection in bodily fluids: clues to transmission. *Clin. Infect. Dis.* **43:**e9–e12.
7. **Bhattacharya, R., G. C. Sahoo, M. K. Nayak, K. Rajendran, P. Dutta, U. Mitra, M. K. Bhattacharya, T. N. Naik, S. K. Bhattacharya, and T. Krishnan.** 2007. Detection of genogroup I and II human picobirnaviruses showing small genomic RNA profile causing acute watery diarrhoea among children in Kolkata, India. *Infect. Genet. Evol.* **7:**229–238.
8. **Biagini, P.** 2004. Human circoviruses. *Vet. Microbiol.* **98:**95–101.
9. **Bidawid, S., J. M. Farber, S. A. Sattar, and S. Hayward.** 2000. Heat inactivation of hepatitis A virus in dairy foods. *J. Food Prot.* **63:**522–528.

10. Boccia, D., J. P. Guthmann, H. Klovstad, N. Hamid, M. Tatay, I. Ciglenecki, J. Y. Nizou, E. Nicand, and P. J. Guerin. 2006. High mortality associated with an outbreak of hepatitis E among displaced persons in Darfur, Sudan. *Clin. Infect. Dis.* **42**:1679–1684.

11. Bofill-Mas, S., N. Albinana-Gimenez, P. Clemente-Casares, A. Hundesa, J. Rodriguez-Manzano, A. Allard, M. Calvo, and R. Girones. 2006. Quantification and stability of human adenoviruses and polyomavirus JCPyV in wastewater matrices. *Appl. Environ. Microbiol.* **72**:7894–7896.

12. Bofill-Mas, S., and R. Girones. 2003. Role of the environment in the transmission of JC virus. *J. Neurovirol.* **9**(Suppl. 1):54–58.

12a. Bouwknegt, M., F. Lodder-Verschoor, W. H. Van der Poel, S. A. Rutjes, and A. M. De Roda Husman. 2007. Hepatitis E virus RNA in commercial porcine livers in The Netherlands. *J. Food Prot.* **70**:2889–2895.

13. Boxman, I. L., J. J. Tilburg, N. A. Te Loeke, H. Vennema, K. Jonker, E. de Boer, and M. Koopmans. 2006. Detection of noroviruses in shellfish in The Netherlands. *Int. J. Food Microbiol.* **108**:391–396.

14. Burke, D. S., and T. P. Monath. 2001. Flaviviruses, p. 1043–1126. *In* D. M. Knipe and P. M. Howley (ed.), *Fields Virology*, 4th ed. Lippincott Williams & Wilkins Philadelphia, PA.

15. Calder, L., G. Simmons, C. Thornley, P. Taylor, K. Pritchard, G. Greening, and J. Bishop. 2003. An outbreak of hepatitis A associated with consumption of raw blueberries. *Epidemiol. Infect.* **131**:745–751.

16. Camitz, M., and F. Liljeros. 2006. The effect of travel restrictions on the spread of a moderately contagious disease. *BMC Med.* **4**:32.

17. Campo, M. S. 1997. Bovine papillomavirus and cancer. *Vet. J.* **154**:175–188.

18. Capua, I., and S. Marangon. 2006. Control and prevention of avian influenza in an evolving scenario. *Vaccine* **25**:5645–5652.

19. Centers for Disease Control and Prevention. 2002. Possible West Nile virus transmission to an infant through breast-feeding—Michigan, 2002. *Morb. Mortal. Wkly. Rep.* **51**:877–878.

20. Chambers, G., V. A. Ellsmore, P. M. O'Brien, S. W. Reid, S. Love, M. S. Campo, and L. Nasir. 2003. Association of bovine papillomavirus with the equine sarcoid. *J. Gen. Virol.* **84**:1055–1062.

21. Chen, W. J., J. Y. Yang, J. H. Lin, C. S. Fann, V. Osyetrov, C. C. King, Y. M. Chen, H. L. Chang, H. W. Kuo, F. Liao, and M. S. Ho. 2006. Nasopharyngeal shedding of severe acute respiratory syndrome-associated coronavirus is associated with genetic polymorphisms. *Clin. Infect. Dis.* **42**:1561–1569.

22. Cheng, V. C., I. F. Hung, B. S. Tang, C. M. Chu, M. M. Wong, K. H. Chan, A. K. Wu, D. M. Tse, K. S. Chan, B. J. Zheng, J. S. Peiris, J. J. Sung, and K. Y. Yuen. 2004. Viral replication in the nasopharynx is associated with diarrhea in patients with severe acute respiratory syndrome. *Clin. Infect. Dis.* **38**:467–475.

23. Chiu, Y. C., K. L. Wu, Y. P. Chou, T. V. Fong, T. L. Tsai, C. M. Kuo, C. H. Kuo, K. W. Chiu, J. W. Liu, H. L. Eng, B. Jawan, Y. F. Cheng, and C. L. Chen. 2004. Diarrhea in medical care workers with severe acute respiratory syndrome. *J. Clin. Gastroenterol.* **38**:880–882.

24. Chomel, B. B., A. Belotto, and F. X. Meslin. 2007. Wildlife, exotic pets, and emerging zoonoses. *Emerg. Infect. Dis.* **13**:6–11.

25. Chua, K. B. 2003. Nipah virus outbreak in Malaysia. *J. Clin. Virol.* **26**:265–275.

26. Chua, K. B., W. J. Bellini, P. A. Rota, B. H. Harcourt, A. Tamin, S. K. Lam, T. G. Ksiazek, P. E. Rollin, S. R. Zaki, W. Shieh, C. S. Goldsmith, D. J. Gubler, J. T. Roehrig, B. Eaton, A. R. Gould, J. Olson, H. Field, P. Daniels, A. E. Ling, C. J. Peters, L. J. Anderson, and B. W. Mahy. 2000. Nipah virus: a recently emergent deadly paramyxovirus. *Science* **288**:1432–1435.

27. Chua, K. B., B. H. Chua, and C. W. Wang. 2002. Anthropogenic deforestation, El Nino and the emergence of Nipah virus in Malaysia. *Malays. J. Pathol.* **24**:15–21.

28. Chua, K. B., S. K. Lam, K. J. Goh, P. S. Hooi, T. G. Ksiazek, A. Kamarulzaman, J. Olson, and C. T. Tan. 2001. The presence of Nipah virus in respiratory secretions and urine of patients during an outbreak of Nipah virus encephalitis in Malaysia. *J. Infect.* **42**:40–43.

29. Courgnaud, V., M. Muller-Trutwin, and P. Sonigo. 2004. Evolution and virulence of primate lentiviruses. *Med. Sci.* (Paris) **20**:448–452. (In French.)

30. Cyranoski, D., and A. Abbott. 2003. Apartment complex holds clues to pandemic potential of SARS. *Nature* **423**:3–4.

31. de Haan, T. R., M. F. Beersma, D. Oepkes, E. P. de Jong, A. C. Kroes, and F. J. Walther. 2007. Parvovirus B19 infection in pregnancy: maternal and fetal viral load measurements related to clinical parameters. *Prenat. Diagn.* **27**:46–50.

32. Duan, S. M., X. S. Zhao, R. F. Wen, J. J. Huang, G. H. Pi, S. X. Zhang, J. Han, S. L. Bi, L. Ruan, and X. P. Dong. 2003. Stability of SARS coronavirus in human specimens and environment and its sensitivity to heating and UV irradiation. *Biomed. Environ. Sci.* **16**: 246–255.

33. Duizer, E., and M. Koopmans. 2006. Tracking foodborne viruses: lessons from noroviruses, p. 77–110. *In* Y. Motarjemi and M. Adams (ed.), *Emerging Foodborne Pathogens.* Woodland Publishing, Cambridge, United Kingdom.

34. Epstein, J. H., H. E. Field, S. Luby, J. R. Pulliam, and P. Daszak. 2006. Nipah virus: impact, origins, and causes of emergence. *Curr. Infect. Dis. Rep.* **8**:59–65.

35. Fauci, A. S., A. M. Macher, D. L. Longo, H. C. Lane, A. H. Rook, H. Masur, and E. P. Gelmann. 1984. NIH conference. Acquired immunodeficiency syndrome: epidemiologic, clinical, immunologic, and therapeutic considerations. *Ann. Intern. Med.* **100**:92–106.

36. Feagins, A. R., T. Opriessnig, D. K. Guenette, P. G. Halbur, and X. J. Meng. 2007. Detection and characterization of infectious hepatitis E virus from commercial pig livers sold in local grocery stores in the USA. *J. Gen. Virol.* **88**:912–917.

37. Gentsch, J. R., P. A. Woods, M. Ramachandran, B. K. Das, J. P. Leite, A. Alfieri, R. Kumar, M. K. Bhan, and R. I. Glass. 1996. Review of G and P typing results from a global collection of rotavirus strains: implications for vaccine development. *J. Infect. Dis.* **174** (Suppl. 1):S30–S36.

38. Greening, G. E. 2006. Human and animal viruses in food (including taxonomy of enteric viruses), p. 5–42. *In* S. Goyal (ed.), *Viruses in Foods.* Springer, New York, NY.

39. Gritsun, T. S., V. A. Lashkevich, and E. A. Gould. 2003. Tick-borne encephalitis. *Antiviral Res.* **57**:129–146.

40. Guthmann, J. P., H. Klovstad, D. Boccia, N. Hamid, L. Pinoges, J. Y. Nizou, M. Tatay, F. Diaz, A. Moren, R. F. Grais, I. Ciglenecki, E. Nicand, and P. J. Guerin. 2006. A large outbreak of hepatitis E among a displaced population in Darfur, Sudan, 2004: the role of water treatment methods. *Clin. Infect. Dis.* **42:**1685–1691.

41. Haglund, M. 2002. Occurrence of TBE in areas previously considered being non-endemic: Scandinavian data generate an international study by the International Scientific Working Group for TBE (ISW-TBE). *Int. J. Med. Microbiol.* **291** (Suppl. 33):50–54.

42. Haglund, M., M. Forsgren, G. Lindh, and L. Lindquist. 1996. A 10-year follow-up study of tick-borne encephalitis in the Stockholm area and a review of the literature: need for a vaccination strategy. *Scand. J. Infect. Dis.* **28:**217–224.

43. Han, M. G., D. S. Cheon, X. Zhang, and L. J. Saif. 2006. Cross-protection against a human enteric coronavirus and a virulent bovine enteric coronavirus in gnotobiotic calves. *J. Virol.* **80:**12350–12356.

44. Harding, J. C. 2004. The clinical expression and emergence of porcine circovirus 2. *Vet. Microbiol.* **98:**131–135.

45. Hecht, R., A. Alban, K. Taylor, S. Post, N. B. Andersen, and R. Schwarz. 2006. Putting it together: AIDS and the millennium development goals. *PLoS Med.* **3:**e455.

46. Hedberg, C. W., S. J. Smith, E. Kirkland, V. Radke, T. F. Jones, and C. A. Selman. 2006. Systematic environmental evaluations to identify food safety differences between outbreak and nonoutbreak restaurants. *J. Food Prot.* **69:**2697–2702.

47. Hino, S., and H. Miyata. 2007. Torque teno virus (TTV): current status. *Rev. Med. Virol.* **17:**45–57.

48. Ho, H. T., M. S. Chang, T. Y. Wei, W. S. Hsieh, C. C. Hung, H. M. Yang, and Y. T. Lu. 2006. Colonization of severe acute respiratory syndrome-associated coronavirus among health-care workers screened by nasopharyngeal swab. *Chest* **129:**95–101.

49. Holmes, K. V. 2001. Coronaviruses, p. 1187–1204. *In* D. M. Knipe and P. M. Howley (ed.), *Fields Virology.* Lippincott Williams & Wilkins, Philadelphia, PA.

50. Hsu, V. P., M. J. Hossain, U. D. Parashar, M. M. Ali, T. G. Ksiazek, I. Kuzmin, M. Niezgoda, C. Rupprecht, J. Bresee, and R. F. Breiman. 2004. Nipah virus encephalitis reemergence, Bangladesh. *Emerg. Infect. Dis.* **10:**2082–2087.

51. Jacobson, E. R., P. E. Ginn, J. M. Troutman, L. Farina, L. Stark, K. Klenk, K. L. Burkhalter, and N. Komar. 2005. West Nile virus infection in farmed American alligators (*Alligator mississippiensis*) in Florida. *J. Wildl. Dis.* **41:**96–106.

52. Kampf, G., J. Steinmann, and H. Rabenau. 2007. Suitability of vaccinia virus and bovine viral diarrhea virus (BVDV) for determining activities of three commonly-used alcohol-based hand rubs against enveloped viruses. *BMC Infect. Dis.* **7:**5.

53. Keawcharoen, J., K. Oraveerakul, T. Kuiken, R. A. Fouchier, A. Amonsin, S. Payungporn, S. Noppornpanth, S. Wattanodorn, A. Theamboonlers, R. Tantilertcharoen, R. Pattanarangsan, N. Arya, P. Ratanakorn, D. M. Osterhaus, and Y. Poovorawan. 2004. Avian influenza H5N1 in tigers and leopards. *Emerg. Infect. Dis.* **10:**2189–2191.

54. Kerbo, N., I. Donchenko, K. Kutsar, and V. Vasilenko. 2005. Tickborne encephalitis outbreak in Estonia linked to raw goat milk, May–June 2005. *Eur. Surveill.* **10:**E050623.2.

55. Klenk, K., J. Snow, K. Morgan, R. Bowen, M. Stephens, F. Foster, P. Gordy, S. Beckett, N. Komar, D. Gubler, and M. Bunning. 2004. Alligators as West Nile virus amplifiers. *Emerg. Infect. Dis.* **10**:2150–2155.

56. Knowles, W. A. 2006. Discovery and epidemiology of the human polyomaviruses BK virus (BKV) and JC virus (JCV). *Adv. Exp. Med. Biol.* **577**:19–45.

57. Koopmans, M., and E. Duizer. 2004. Foodborne viruses: an emerging problem. *Int. J. Food Microbiol.* **90**:23–41.

58. Koopmans, M., H. Vennema, H. Heersma, E. van Strien, Y. van Duynhoven, D. Brown, M. Reacher, and B. Lopman. 2003. Early identification of common-source foodborne virus outbreaks in Europe. *Emerg. Infect. Dis.* **9**:1136–1142.

59. Koopmans, M., B. Wilbrink, M. Conyn, G. Natrop, H. van der Nat, H. Vennema, A. Meijer, J. van Steenbergen, R. Fouchier, A. Osterhaus, and A. Bosman. 2004. Transmission of H7N7 avian influenza A virus to human beings during a large outbreak in commercial poultry farms in The Netherlands. *Lancet* **363**:587–593.

60. Kuiken, T., G. Rimmelzwaan, D. van Riel, G. van Amerongen, M. Baars, R. Fouchier, and A. Osterhaus. 2004. Avian H5N1 influenza in cats. *Science* **306**:241.

61. Labuda, M., E. Eleckova, M. Lickova, and A. Sabo. 2002. Tick-borne encephalitis virus foci in Slovakia. *Int. J. Med. Microbiol.* **291** (Suppl. 33):43–47.

62. Lai, M. M. C., and K. V. Holmes. 2001. *Coronaviridae*: the viruses and their replication, p. 1163–1186. *In* D. M., Knipe and P. M Howley (ed.), *Fields Virology*. Lippincott Williams & Wilkins, Philadelphia, PA.

63. Laird, A. R., V. Ibarra, G. Ruiz-Palacios, M. L. Guerrero, R. I. Glass, and J. R. Gentsch. 2003. Unexpected detection of animal VP7 genes among common rotavirus strains isolated from children in Mexico. *J. Clin. Microbiol.* **41**:4400–4403.

64. Lamarre, A., and P. J. Talbot. 1989. Effect of pH and temperature on the infectivity of human coronavirus 229E. *Can. J. Microbiol.* **35**:972–974.

65. Lau, S. K., P. C. Woo, B. H. Wong, H. W. Tsoi, G. K. Woo, R. W. Poon, K. H. Chan, W. I. Wei, J. S. Peiris, and K. Y. Yuen. 2004. Detection of severe acute respiratory syndrome (SARS) coronavirus nucleocapsid protein in SARS patients by enzyme-linked immunosorbent assay. *J. Clin. Microbiol.* **42**:2884–2889.

66. Le Guyader, F. S., F. Bon, D. DeMedici, S. Parnaudeau, A. Bertone, S. Crudeli, A. Doyle, M. Zidane, E. Suffredini, E. Kohli, F. Maddalo, M. Monini, A. Gallay, M. Pommepuy, P. Pothier, and F. M. Ruggeri. 2006. Detection of multiple noroviruses associated with an international gastroenteritis outbreak linked to oyster consumption. *J. Clin. Microbiol.* **44**:3878–3882.

67. Leung, W. K., K. F. To, P. K. Chan, H. L. Chan, A. K. Wu, N. Lee, K. Y. Yuen, and J. J. Sung. 2003. Enteric involvement of severe acute respiratory syndrome-associated coronavirus infection. *Gastroenterology* **125**:1011–1017.

68. Li, T. C., K. Chijiwa, N. Sera, T. Ishibashi, Y. Etoh, Y. Shinohara, Y. Kurata, M. Ishida, S. Sakamoto, N. Takeda, and T. Miyamura. 2005. Hepatitis E virus transmission from wild boar meat. *Emerg. Infect. Dis.* **11**:1958–1960.

69. Liu, W., F. Tang, A. Fontanet, L. Zhan, Q. M. Zhao, P. H. Zhang, X. M. Wu, S. Q. Zuo, L. Baril, A. Vabret, Z. T. Xin, Y. M. Shao, H. Yang, and W. C. Cao. 2004. Long-term SARS coronavirus excretion from patient cohort, China. *Emerg. Infect. Dis.* **10**:1841–1843.

70. Luby, S. P., M. Rahman, M. J. Hossain, L. S. Blum, M. M. Husain, E. Gurley, R. Khan, B. N. Ahmed, S. Rahman, N. Nahar, E. Kenah, J. A. Comer, and T. G. Ksiazek. 2006. Foodborne transmission of Nipah virus, Bangladesh. *Emerg. Infect. Dis.* **12:**1888–1894.

71. Ludert, J. E., M. Hidalgo, F. Gil, and F. Liprandi. 1991. Identification in porcine faeces of a novel virus with a bisegmented double stranded RNA genome. *Arch. Virol.* **117:**97–107.

72. Maeda, Y., T. Shibahara, Y. Wada, K. Kadota, T. Kanno, I. Uchida, and S. Hatama. 2007. An outbreak of teat papillomatosis in cattle caused by bovine papilloma virus (BPV) type 6 and unclassified BPVs. *Vet. Microbiol.* **121:**242–248.

73. Magar, R., and R. Larochelle. 2004. Evaluation of the presence of porcine reproductive and respiratory syndrome virus in pig meat and experimental transmission following oral exposure. *Can. J. Vet. Res.* **68:**259–266.

74. Maines, T. R., X. H. Lu, S. M. Erb, L. Edwards, J. Guarner, P. W. Greer, D. C. Nguyen, K. J. Szretter, L. M. Chen, P. Thawatsupha, M. Chittaganpitch, S. Waicharoen, D. T. Nguyen, T. Nguyen, H. H. Nguyen, J. H. Kim, L. T. Hoang, C. Kang, L. S. Phuong, W. Lim, S. Zaki, R. O. Donis, N. J. Cox, J. M. Katz, and T. M. Tumpey. 2005. Avian influenza (H5N1) viruses isolated from humans in Asia in 2004 exhibit increased virulence in mammals. *J. Virol.* **79:**11788–11800.

75. Masachessi, G., L. C. Martinez, M. O. Giordano, P. A. Barril, B. M. Isa, L. Ferreyra, D. Villareal, M. Carello, C. Asis, and S. V. Nates. 2007. Picobirnavirus (PBV) natural hosts in captivity and virus excretion pattern in infected animals. *Arch. Virol.* **152:**989–998.

76. Mast, E. E. 2004. Mother-to-infant hepatitis C virus transmission and breastfeeding. *Adv. Exp. Med. Biol.* **554:**211–216.

77. Matrosovich, M., T. Matrosovich, J. Uhlendorff, W. Garten, and H. D. Klenk. 2007. Avian-virus-like receptor specificity of the hemagglutinin impedes influenza virus replication in cultures of human airway epithelium. *Virology* **361:**384–390.

78. McKinney, K. R., Y. Y. Gong, and T. G. Lewis. 2006. Environmental transmission of SARS at Amoy Gardens. *J. Environ. Health* **68:**26–30; quiz, 51–52.

79. Meng, X. J., R. H. Purcell, P. G. Halbur, J. R. Lehman, D. M. Webb, T. S. Tsareva, J. S. Haynes, B. J. Thacker, and S. U. Emerson. 1997. A novel virus in swine is closely related to the human hepatitis E virus. *Proc. Natl. Acad. Sci. USA* **94:**9860–9865.

80. Middleton, D. J., H. A. Westbury, C. J. Morrissy, B. M. van der Heide, G. M. Russell, M. A. Braun, and A. D. Hyatt. 2002. Experimental Nipah virus infection in pigs and cats. *J. Comp. Pathol.* **126:**124–136.

81. Monteny, M., H. G. Niesters, H. A. Moll, and M. Y. Berger. 2007. Human bocavirus in febrile children, The Netherlands. *Emerg. Infect. Dis.* **13:**180–182.

82. Morse, S. S. 1995. Factors in the emergence of infectious diseases. *Emerg. Infect. Dis.* **1:**7–15.

83. Negrete, O. A., E. L. Levroney, H. C. Aguilar, A. Bertolotti-Ciarlet, R. Nazarian, S. Tajyar, and B. Lee. 2005. EphrinB2 is the entry receptor for Nipah virus, an emergent deadly paramyxovirus. *Nature* **436:**401–405.

84. Nicholls, M., M. Bruce, and J. Thomas. 2003. Management of hepatitis A in a food handler at a London secondary school. *Commun. Dis. Public Health* **6:**26–29.

85. Olsen, B., V. J. Munster, A. Wallensten, J. Waldenstrom, A. D. Osterhaus, and R. A. Fouchier. 2006. Global patterns of influenza A virus in wild birds. *Science* 312:384–388.

86. Ozawa, Y., B. L. Ong, and S. H. An. 2001. Traceback systems used during recent epizootics in Asia. *Rev. Sci. Technol.* 20:605–613.

87. Panina, G. F., A. Civardi, P. Cordioli, I. Massirio, F. Scatozza, P. Baldini, and F. Palmia. 1992. Survival of hog cholera virus (HCV) in sausage meat products (Italian salami). *Int. J. Food Microbiol.* 17:19–25.

88. Panina, G. F., A. Civardi, I. Massirio, F. Scatozza, P. Baldini, and F. Palmia. 1989. Survival of foot-and-mouth disease virus in sausage meat products (Italian salami). *Int. J. Food Microbiol.* 8:141–148.

89. Parashar, U. D., L. Dow, R. L. Fankhauser, C. D. Humphrey, J. Miller, T. Ando, K. S. Williams, C. R. Eddy, J. S. Noel, T. Ingram, J. S. Bresee, S. S. Monroe, and R. I. Glass. 1998. An outbreak of viral gastroenteritis associated with consumption of sandwiches: implications for the control of transmission by food handlers. *Epidemiol. Infect.* 121:615–621.

90. Parry, J. 2003. WHO issues global alert on respiratory syndrome. *Br. Med. J.* 326:615.

91. Pealer, L. N., A. A. Marfin, L. R. Petersen, R. S. Lanciotti, P. L. Page, S. L. Stramer, M. G. Stobierski, K. Signs, B. Newman, H. Kapoor, J. L. Goodman, and M. E. Chamberland. 2003. Transmission of West Nile virus through blood transfusion in the United States in 2002. *N. Engl. J. Med.* 349:1236–1245.

92. Phipps, L. P., R. E. Gough, V. Ceeraz, W. J. Cox, and I. H. Brown. 2007. Detection of West Nile virus in the tissues of specific pathogen free chickens and serological response to laboratory infection: a comparative study. *Avian Pathol.* 36:301–305.

93. Polywka, S., M. Schroter, H. H. Feucht, B. Zollner, and R. Laufs. 1999. Low risk of vertical transmission of hepatitis C virus by breast milk. *Clin. Infect. Dis.* 29:1327–1329.

94. Promkuntod, N., C. Antarasena, P. Prommuang, and P. Prommuang. 2006. Isolation of avian influenza virus A subtype H5N1 from internal contents (albumen and allantoic fluid) of Japanese quail (*Coturnix coturnix japonica*) eggs and oviduct during a natural outbreak. *Ann. N. Y. Acad. Sci.* 1081:171–173.

95. Rabenau, H. F., J. Cinatl, B. Morgenstern, G. Bauer, W. Preiser, and H. W. Doerr. 2005. Stability and inactivation of SARS coronavirus. *Med. Microbiol. Immunol.* 194:1–6.

96. Rahman, M., J. Matthijnssens, X. Yang, T. Delbeke, I. Arijs, K. Taniguchi, M. Iturriza-Gomara, N. Iftekharuddin, T. Azim, and M. Van Ranst. 2007. Evolutionary history and global spread of the emerging g12 human rotaviruses. *J. Virol.* 81:2382–2390.

97. Raoult, D., S. Audic, C. Robert, C. Abergel, P. Renesto, H. Ogata, B. La Scola, M. Suzan, and J. M. Claverie. 2004. The 1.2-megabase genome sequence of Mimivirus. *Science* 306:1344–1350.

98. Reed, K. D., J. W. Melski, M. B. Graham, R. L. Regnery, M. J. Sotir, M. V. Wegner, J. J. Kazmierczak, E. J. Stratman, Y. Li, J. A. Fairley, G. R. Swain, V. A. Olson, E. K. Sargent, S. C. Kehl, M. A. Frace, R. Kline, S. L. Foldy, J. P. Davis, and I. K. Damon. 2004. The detection of monkeypox in humans in the Western Hemisphere. *N. Engl. J. Med.* 350:342–350.

99. Reynes, J. M., D. Counor, S. Ong, C. Faure, V. Seng, S. Molia, J. Walston, M. C. Georges-Courbot, V. Deubel, and J. L. Sarthou. 2005. Nipah virus in Lyle's flying foxes, Cambodia. *Emerg. Infect. Dis.* **11**:1042–1047.

100. Ricketts, M. N. 2004. Public health and the BSE epidemic. *Curr. Top. Microbiol. Immunol.* **284**:99–119.

101. Rockx, B., T. Sheahan, E. Donaldson, J. Harkema, A. Sims, M. Heise, R. Pickles, M. Cameron, D. Kelvin, and R. Baric. 2007. Synthetic reconstruction of zoonotic and early human severe acute respiratory syndrome coronavirus isolates that produce fatal disease in aged mice. *J. Virol.* **81**:7410–7423.

102. Roden, R. B., D. R. Lowy, and J. T. Schiller. 1997. Papillomavirus is resistant to desiccation. *J. Infect. Dis.* **176**:1076–1079.

103. Romalde, J. L., E. Area, G. Sanchez, C. Ribao, I. Torrado, X. Abad, R. M. Pinto, J. L. Barja, and A. Bosch. 2002. Prevalence of enterovirus and hepatitis A virus in bivalve molluscs from Galicia (NW Spain): inadequacy of the EU standards of microbiological quality. *Int. J. Food Microbiol.* **74**:119–130.

104. Saknimit, M., I. Inatsuki, Y. Sugiyama, and K. Yagami. 1988. Virucidal efficacy of physico-chemical treatments against coronaviruses and parvoviruses of laboratory animals. *Jikken Dobutsu* **37**:341–345.

105. Sbrana, E., J. H. Tonry, S. Y. Xiao, A. P. da Rosa, S. Higgs, and R. B. Tesh. 2005. Oral transmission of West Nile virus in a hamster model. *Am. J. Trop. Med. Hyg.* **72**:325–329.

106. Schroder, H., J. Marrugat, M. Covas, R. Elosua, A. Pena, T. Weinbrenner, M. Fito, M. A. Vidal, and R. Masia. 2004. Population dietary habits and physical activity modification with age. *Eur. J. Clin. Nutr.* **58**:302–311.

107. Schwegmann-Wessels, C., and G. Herrler. 2006. Sialic acids as receptor determinants for coronaviruses. *Glycoconj. J.* **23**:51–58.

108. Segales, J., G. M. Allan, and M. Domingo. 2005. Porcine circovirus diseases. *Anim. Health Res. Rev.* **6**:119–142.

109. Selvey, L. A., R. M. Wells, J. G. McCormack, A. J. Ansford, K. Murray, R. J. Rogers, P. S. Lavercombe, P. Selleck, and J. W. Sheridan. 1995. Infection of humans and horses by a newly described morbillivirus. *Med. J. Aust.* **162**:642–645.

110. Shi, Z., and Z. Hu. A review of studies on animal reservoirs of the SARS coronavirus. *Virus Res.*, in press.

111. Shibata, I., Y. Okuda, K. Kitajima, and T. Asai. 2006. Shedding of porcine circovirus into colostrum of sows. *J. Vet. Med. Ser B* **53**:278–280.

112. Smith, G. J., T. S. Naipospos, T. D. Nguyen, M. D. de Jong, D. Vijaykrishna, T. B. Usman, S. S. Hassan, T. V. Nguyen, T. V. Dao, N. A. Bui, Y. H. Leung, C. L. Cheung, J. M. Rayner, J. X. Zhang, L. J. Zhang, L. L. Poon, K. S. Li, V. C. Nguyen, T. T. Hien, J. Farrar, R. G. Webster, H. Chen, J. S. Peiris, and Y. Guan. 2006. Evolution and adaptation of H5N1 influenza virus in avian and human hosts in Indonesia and Vietnam. *Virology* **350**:258–268.

113. Sokolow, S. H., C. Rand, S. L. Marks, N. L. Drazenovich, E. J. Kather, and J. E. Foley. 2005. Epidemiologic evaluation of diarrhea in dogs in an animal shelter. *Am. J. Vet. Res.* **66**:1018–1024.

114. Songsermn, T., A. Amonsin, R. Jam-on, N. Sae-Heng, N. Meemak, N. Pariyothorn, S. Payungporn, A. Theamboonlers, and Y. Poovorawan. 2006. Avian influenza H5N1 in naturally infected domestic cat. *Emerg. Infect. Dis.* **12**:681–683.

115. Stocco dos Santos, R. C., C. J. Lindsey, O. P. Ferraz, J. R. Pinto, R. S. Mirandola, F. J. Benesi, E. H. Birgel, C. A. Pereira, and W. Becak. 1998. Bovine papillomavirus transmission and chromosomal aberrations: an experimental model. *J. Gen. Virol.* **79**:2127–2135.

116. Stratton, K., D. A. Almario, and M. C. McCormick. 2002. *Immunization Safety Review: SV40 Contamination of Polio Vaccine and Cancer.* National Academies Press, Washington, DC.

117. Swayne, D. E., and J. R. Beck. 2005. Experimental study to determine if low-pathogenicity and high-pathogenicity avian influenza viruses can be present in chicken breast and thigh meat following intranasal virus inoculation. *Avian Dis.* **49**:81–85.

118. Taylor, L. H., S. M. Latham, and M. E. Woolhouse. 2001. Risk factors for human disease emergence. *Philos. Trans. R. Soc. Lond. Ser B.* **356**:983–989.

119. Tei, S., N. Kitajima, S. Ohara, Y. Inoue, M. Miki, T. Yamatani, H. Yamabe, S. Mishiro, and Y. Kinoshita. 2004. Consumption of uncooked deer meat as a risk factor for hepatitis E virus infection: an age- and sex-matched case-control study. *J. Med. Virol.* **74**:67–70.

120. Tei, S., N. Kitajima, K. Takahashi, and S. Mishiro. 2003. Zoonotic transmission of hepatitis E virus from deer to human beings. *Lancet* **362**:371–373.

121. Thiry, E., A. Zicola, D. Addie, H. Egberink, K. Hartmann, H. Lutz, H. Poulet, and M. C. Horzinek. 2007. Highly pathogenic avian influenza H5N1 virus in cats and other carnivores. *Vet. Microbiol.* **122**:25–31.

122. Tonry, J. H., S. Y. Xiao, M. Siirin, H. Chen, A. P. da Rosa, and R. B. Tesh. 2005. Persistent shedding of West Nile virus in urine of experimentally infected hamsters. *Am. J. Trop. Med. Hyg.* **72**:320–324.

123. Tsirimonaki, E., B. W. O'Neil, R. Williams, and M. S. Campo. 2003. Extensive papillomatosis of the bovine upper gastrointestinal tract. *J. Comp. Pathol.* **129**:93–99.

124. Uiprasertkul, M., P. Puthavathana, K. Sangsiriwut, P. Pooruk, K. Srisook, M. Peiris, J. M. Nicholls, K. Chokephaibulkit, N. Vanprapar, and P. Auewarakul. 2005. Influenza A H5N1 replication sites in humans. *Emerg. Infect. Dis.* **11**:1036–1041.

125. van der Hoek, L., K. Pyrc, M. F. Jebbink, W. Vermeulen-Oost, R. J. Berkhout, K. C. Wolthers, P. M. Wertheim-van Dillen, J. Kaandorp, J. Spaargaren, and B. Berkhout. 2004. Identification of a new human coronavirus. *Nat. Med.* **10**:368–373.

126. van der Meulen, K. M., M. B. Pensaert, and H. J. Nauwynck. 2005. West Nile virus in the vertebrate world. *Arch. Virol.* **150**:637–657.

127. Vereta, L. A., V. Z. Skorobrekha, S. P. Nikolaeva, V. I. Aleksandrov, V. I. Tolstonogova, T. A. Zakharycheva, A. P. Red'ko, M. I. Lev, and N. A. Savel'eva. 1991. The transmission of the tick-borne encephalitis virus via cow's milk. *Med. Parazitol.* (Moscow) **1991**:54–56. (In Russian.)

128. Vicente, D., G. Cilla, M. Montes, E.G . Perez-Yarza, and E. Perez-Trallero. 2007. Human bocavirus, a respiratory and enteric virus. *Emerg. Infect. Dis.* **13**:636–637.

129. Wacharapluesadee, S., K. Boongird, S. Wanghongsa, P. Phumesin, and T. Hemachudha. 2006. Drinking bat blood may be hazardous to your health. *Clin. Infect. Dis.* 43:269.

130. Wang, L. F., Z. Shi, S. Zhang, H. Field, P. Daszak, and B. T. Eaton. 2006. Review of bats and SARS. *Emerg. Infect. Dis.* 12:1834–1840.

131. Weiss, R. A., and A. J. McMichael. 2004. Social and environmental risk factors in the emergence of infectious diseases. *Nat. Med.* 10:S70–S76.

132. White, M. K., J. Gordon, K. Reiss, L. Del Valle, S. Croul, A. Giordano, A. Darbinyan, and K. Khalili. 2005. Human polyomaviruses and brain tumors. *Brain Res. Brain Res. Rev.* 50:69–85.

133. Woo, P. C., S. K. Lau, C. M. Chu, K. H. Chan, H. W. Tsoi, Y. Huang, B. H. Wong, R. W. Poon, J. J. Cai, W. K. Luk, L. L. Poon, S. S. Wong, Y. Guan, J. S. Peiris, and K. Y. Yuen. 2005. Characterization and complete genome sequence of a novel coronavirus, coronavirus HKU1, from patients with pneumonia. *J. Virol.* 79:884–895.

134. World Health Organization. 2007. *Viruses in Food: Scientific Advice To Support Risk Management Activities.* World Health Organization, Geneva, Switzerland. www.who.int.

135. Yazaki, Y., H. Mizuo, M. Takahashi, T. Nishizawa, N. Sasaki, Y. Gotanda, and H. Okamoto. 2003. Sporadic acute or fulminant hepatitis E in Hokkaido, Japan, may be food-borne, as suggested by the presence of hepatitis E virus in pig liver as food. *J. Gen. Virol.* 84:2351–2357.

Food-Borne Viruses: Progress and Challenges
Edited by Marion P. G. Koopmans, Dean O. Cliver, and Albert Bosch
© 2008 ASM Press, Washington, DC

Viral Evolution and Its Relevance for Food-Borne Virus Epidemiology

6

Esteban Domingo and Harry Vennema

The adaptation of viruses to changing environments determines the types and numbers of viral particles present in any physical context, be it food or other, and hence the probability of encountering a susceptible host to cause disease (see chapter 8). The stability of virions as physical particles is a reflection of the evolutionary history of the virus. Transmission as free particles to a susceptible host necessitates at least a minimal half-life of their infectivity. This book concerns the multiple factors that define food-borne and waterborne viral illness, the transmissibility of the viral agents, risk assessment, and preventive measures. In this context, the viruses known to be transmissible by food or water are those that have evolved phenotypic traits that represent a danger which we can identify and even in some cases anticipate. Review of data from outbreak reporting has shown that genogroup I noroviruses (NoV) are significantly more often associated with food-borne and waterborne transmission than are the dominant strains that circulate in the community, the genogroup II4 viruses. How could evolution select for food-borne viruses? Whether food-borne transmission influences virus evolution is a second question that is addressed in this chapter. These difficult questions concern virology, epidemiology, and evolutionary biology; their answers require first and foremost some understanding of how viral populations diversify in nature.

ESTEBAN DOMINGO, Centro de Biología Molecular "Severo Ochoa" (CSIC-UAM), Universidad Autónoma de Madrid, Cantoblanco, 28049 Madrid, Spain. HARRY VENNEMA, Laboratory for Infectious Diseases, Centre for Infectious Diseases Control, National Institute for Public Health and the Environment, 3720 BA Bilthoven, The Netherlands.

GENERAL INTRODUCTION TO VIRUS EVOLUTION

Studies of virus evolution include a broad range of topics among which we can distinguish the following (reviewed in reference 16).

- Theories of the origin and early evolution of viruses
- Evolutionary relationships among present-day viruses
- Short-term evolution of viruses within infected host organisms

The origin and early evolution of viruses cannot be approached experimentally. Models of early events have been put forward on the basis of comparisons of genomic sequences of host organisms and their parasites, including viruses. Unavoidably, however, there must be a considerable degree of speculation in the proposals. In very broad terms, models of virus origin can be divided in two groups: (i) models proposing that viruses are remnants of a primitive world of independent "replicons" (small RNA or RNA-like molecules which acquired the capacity to perpetuate some minimal information), either preceding or coexisting with early cellular life; and (ii) models proposing that viruses are modern derivatives of a world of differentiated cells. Genetic elements that were part of cellular genomes could have acquired autonomous replication and an extracellular phase in their life cycle, or the entire cells could have gradually become more dependent on other cells, with gradual loss of genetic information, leading eventually to the genomic forms that preceded the complex DNA viruses we know today. A widely accepted view is that functional modules have been shared by cells and viruses and that viruses have played a major role in the development of life on Earth (16, 29, 65). These events, however remote they might seem, are not alien to those that determine the adaptability of present-day viruses, since they must have involved mutation and recombination (discussed in the next sections). The mimiviruses (or microbe-mimicking virus) contain a 1,200-kbp double-stranded DNA genome wrapped into a capsid that is 300 nm in diameter, about 10 times the size of the common viruses with a capsid of icosahedral symmetry. The existence of such large viruses supports the theory that at least some viruses may be regressed forms of cells. In fact, the two main theories of virus origins are not mutually exclusive. While some simple RNA genetic elements may be remnants of a primitive RNA world, the complex DNA viruses may have evolved from and within organized cellular ensembles.

The evolutionary relationships among present-day viruses can be experimentally approached by comparing partial or complete genomic nucleotide and protein sequences, either of different viruses or of different isolates in the same virus group. Important progress has been achieved through new methodologies for virus isolation, selective amplification of viral nucleic acids, automated nucleotide sequencing procedures, and methods of aligning, comparing, and interpreting nucleotide and deduced amino acid sequences; this area of research can be regarded as part of bioinformatics and has been termed "virogenomics" (18). Although evolutionary relationships among viruses can be (and have been) established on the basis of structural comparisons of the viral particles or other phenotypic traits, the ease of obtaining nucleotide sequences of viruses, a trend that will continue in coming years (thanks in part to pyrosequencing and related methodology), means that viral genomics is the major method used for investigating viral evolution. Phylogenetic procedures, developed largely in the field of general population genetics, are routinely applied to the comparison of viral nucleotide and protein sequences (for reviews of these techniques, see references 16, 50, and 58).

The two most common food-borne viruses are hepatitis A virus (HAV) and NoV. Human HAV, together with simian HAV, constitutes the genus *Hepatovirus*, family *Picornaviridae* (see chapter 3). Another tentative species in this genus is avian encephalomyelitis-like virus. HAV can be divided into genotypes, of which IA and IB are the most common. The HAV strains have more than 80% nucleotide sequence similarity; the most highly conserved regions are the 5′ noncoding region and the nonstructural-protein coding region. The structural-protein coding region is most suitable for strain differentiation.

The genus *Norovirus* belongs to the family *Caliciviridae*. The family includes four distinct genera (*Norovirus*, *Lagovirus*, *Sapovirus*, and *Vesivirus*) and a fifth one containing some unassigned viruses (e.g., a bovine enteric calicivirus strain). The most common human NoV strains belong to two genogroups, which are further subdivided into genotypes (see chapter 2). NoV strains belonging to one genotype have at least 80% amino acid sequence homology in the capsid protein. Highly conserved regions in the genome such as the polymerase gene can be up to 15% different in their nucleotide sequence among NoV belonging to the same genotype (36, 66, 68, 73).

Genes encoding nonstructural proteins of viruses are generally more highly conserved than are genes encoding structural proteins such as capsid proteins of naked viruses and surface glycoproteins of enveloped viruses. Establishing phylogenetic relationships by using variable genes is adequate to trace the origins of a disease outbreak to a specific source of virus, for example contaminated food as the only origin of multiple disease episodes. Comparison of the shape and branch lengths of phylogenetic trees can be indicative of the mode and tempo of the evolution of a virus. Furthermore, a discordant phylogenetic position of two different genes of the same virus is indicative of recombination, a frequent occurrence for DNA viruses and many RNA viruses (16, 43, 50, 58) (see the next section).

The mechanism most directly relevant to viral pathogenesis and to the topics covered in this volume is the short-term evolution of viruses within infected hosts. These are microevolutionary events that take place within short time intervals (days or weeks) and that may or may not impact the middle- and long-term evolution of the virus. For example, short-term evolution may have provided stability for viral particles to survive within the gastrointestinal tract. To describe the events involved in evolution within the host, we briefly discuss the molecular mechanism of genetic variation, as well as the dynamics of viral populations in vivo, which are also supported by studies with model systems in cell culture.

MOLECULAR MECHANISMS OF GENETIC VARIATION OF VIRUSES

Viruses use the same mechanisms of genetic variation as any form of life known to us: mutation, recombination, and genome-segment reassortment. Mutation is the most universal form of genetic variation of viruses. It affects all viral and subviral particles known, because it originates in inaccurate interactions between template bases and nucleotide substrates at the active site of the viral polymerases. Recombination, or the formation of a genome which results from the covalent linkage of two (or more) parental genomes, appears to vary in occurrence among different types of viruses, being more frequent among DNA viruses, retroviruses, and plus-strand riboviruses. Segment reassortment occurs in viruses with a segmented genome, typically in the influenza viruses but also, more relevant for the enteric virus field, in the rotaviruses. The rotavirus genome includes 11 segments of double-stranded RNA that can be separated by polyacrylamide gel electrophoresis. Reassortment can give rise to distinct electropherograms, which in some cases have been useful to trace the spread of the virus through a human population. Thus, both mutation and the reassortment of genome segments can be useful markers for studies of the epidemiology of viral disease. While mutation is inherent to the replication of any genome in an infected cell, recombination and segment reassortment necessitate coinfection of the same cell by at least two parental viruses. For recombination or reassortment to be detected, the parental viruses must be distinguishable genetically. Mutation and recombination are closely related to the enzymology of DNA and RNA replication, and both play distinct roles in evolution, as summarized in the next section.

Mutation and Its Role in Virus Evolution

Mutation of viral RNA and DNA can occur through chemical damage to the template or as a result of incorporation of an incorrect (noncomplementary) nucleotide during nucleic acid synthesis. Misincorporation is influenced by the proportion of unusual tautomeric forms of the standard nucleotides (forms that differ in the hydrogen bonds which they can establish with other bases), occurrence of alternative "wobble" base pairs, and effects of base stacking in the template molecule (reviewed in reference 30).

Once an incorrect base is incorporated into the growing nucleotide chain, it can be excised—and the correct nucleotide introduced—provided that the polymerase includes a proofreading-repair activity, generally in the form of a 3′-to-5′ exonuclease. Many cellular and some viral DNA polymerases (those of complex viruses such as the adenoviruses, herpesviruses, and poxviruses) include such an exonuclease activity. Most viral RNA-dependent RNA polymerases and retrotranscriptases lack a proofreading-repair activity, as evidenced by both biochemical and structural studies (28, 61). Some coronaviruses include an exonuclease activity (Exo N in nsp14 of the severe acute respiratory syndrome [SARS] coronavirus) that is essential for RNA replication, although its possible role in RNA copying fidelity has not been demonstrated (46). The absence (or low efficiency) of proofreading-repair activity contributes to the elevated mutation rates of riboviruses and retroviruses, estimated at 10^{-3} to 10^{-5} substitution introduced per nucleotide copied (4, 21), while the rate of evolution in nature is in the range of 10^{-2} to 10^{-4} substitution per nucleotide per year (Table 1).

Cells include correction mechanisms for postreplicative mismatches (i.e., unpaired bases that have escaped proofreading-repair) (30); these mechanisms act on double-stranded DNA but do not act (or act inefficiently) on double-stranded RNA or RNA-DNA hybrids. This, again, would favor higher mutation rates for RNA viruses (or any virus that uses RNA as a replicative intermediate in its replication cycle) than for DNA viruses. To substantiate these predicted differences, determinations of mutation rates for DNA viruses of different groups are needed as a counterpart of the several determinations for RNA viruses. The available measurements suggest the existence of large variations in both genetic heterogeneity and rates of evolution in nature, both for RNA and DNA viruses.

When studying the population structure of the most common food-borne virus, NoV, it is clear that all of these NoV strains evolved from a common ancestor a long time ago. In contrast, HAV has been relatively stable despite having an RNA genome; this shows that genetic flexibility of RNA genomes is not necessarily reflected in continuing or rapid evolution.

Table 1 Mutation frequencies and rates of evolution for some picornaviruses[a]

Virus	Mutation frequency and evolution rate[b]	Method
Poliovirus	$(2.0 \pm 1.2) \times 10^{-3} - (4.1 \pm 0.6) \times 10^{-3}$ s/nt	Biochemical. Frequency of resistance to T_1 RNase at G residues.
	$(2.1 \pm 1.9) \times 10^{-4} - (6.5 \pm 6.3) \times 10^{-4}$ s/nt	Genetic. Frequency of transition from drug dependence to drug independence.
	$10^{-5} - 10^{-3}$ s/nt	Genetic. Frequency of mutants resistant to monoclonal antibodies.
	$<10^{-6}$ s/nt	Biochemical. Frequency of mutations in virus from individual plaques.
	2.5×10^{-6} s/nt	Genetic. Frequency of reversion of an amber mutant.
	2×10^{-5} s/nt	Genetic. Frequency of reversion of a point mutant.
	$6.9 \times 10^{-3} - 1.4 \times 10^{-2}$ s/nt/y	Genetic. Average rate of accumulation of mutations over the genome.
Coxsackievirus A9	1×10^{-4} s/nt	Genetic. Frequency of transition from drug dependence to drug independence.
Coxsackievirus A24	3.7×10^{-3} s/nt/y	Genetic. Average rate of accumulation of mutations at gene 3°C.
Enterovirus 70	5×10^{-3} s/nt/y	Genetic. Average rate of accumulation of mutations at gene VP1.
Enterovirus 71	1.3×10^{-2} s/nt/y	Genetic. Average rate of accumulation of mutations at gene VP1.
Hepatitis A virus	$1.4 \times 10^{-4} - 7 \times 10^{-4}$ s/nt	Biochemical. Sequencing of genomes from clonal populations.

[a]Based on references 16, 17, 21, and 59.
[b]s/nt, substitutions per nucleotide; s/nt/y, substitutions per nucleotide per year.

Since most food-borne viruses are RNA viruses, factors determining mutation rates for DNA viruses are not discussed further. A comparison of mutation rates for RNA viruses and DNA viruses and a compilation of several values can be found in reference 16. Studies of several cellular and viral polymerases have demonstrated that amino acid replacements at some key residues of the enzymes can affect the mutation rate (3, 16, 30, 45, 55, 64). This, in turn, suggests that mutation rates can themselves be objects of selection (22), implying that viral systems may, in their prolonged evolution-ary history, have been able to adjust their mutation rates in response to envi-ronmental and replicative requirements. It has been speculated that high mutation rates for RNA viruses are simply a consequence of rapid RNA replication. Current evidence does not support this speculation because high mutation rates have proven essential for the viruses to adapt to complex envi-ronments, not all RNA viruses have a lifestyle that requires rapid replication, and enzymological evidence indicates that multiple determinants (not only the rate of nucleotide incorporation) are involved in template-copying fidelity (further arguments are given in reference 19).

A distinction should be made between mutation rate, mutation frequency, and rate of accumulation (or fixation) of mutations. The distinction clarifies the population structure of viral populations in that a virus may show elevated mutation frequencies (relevant to population heterogeneity, short-term evolution, and viral pathogenesis) and a low rate of accumulation of muta-tions or a low rate of evolution (relevant to the long-term diversification of viruses in nature). Despite a coincidental range of values (see the values in Table 1), mutation frequencies and rates of evolution have clearly different meanings. Unfortunately, in the population genetics literature, the rate of evolution is often referred to as the mutation rate, obscuring basic scientific concepts underlying the two measurements.

Recombination and Its Role in Virus Evolution

Several forms of recombination in viral genomes have been characterized. Recombination is defined as homologous when there is extensive nucleotide sequence identity between the two parental genomes around the crossover site (the junction between the two parental sequences in the recombinant genome) and nonhomologous when no significant nucleotide sequence iden-tity around the crossover site is observed. A second distinction is made between replicative and nonreplicative recombination, depending on whether replication of the viral genome is required or not required for recombination to occur. Recombination events can be intergenomic (involving two or more different genomes) or intragenomic (involving the same genome). The occurrence of internal deletions, such as in the generation of defective inter-fering (DI) RNA and DI particles, is often mediated by intragenomic recombination.

Recombination is a widespread mechanism of genetic variation of both DNA and RNA viruses, although some types of recombination appear to be absent or highly restricted in some virus groups. For example, homologous replicative intergenomic recombination is very frequent among some groups of plus-strand RNA viruses (including picornaviruses, caliciviruses, and coronaviruses) but appears to be infrequent among minus-strand RNA viruses (including the rhabdoviruses). However, several minus-strand RNA viruses produce DI RNAs at high frequency in the course of passages at high multiplicities of infection. For reviews of the different types of genetic recombination in viruses, see references 7, 16, 33, and 48.

Recombination probably plays two disparate roles in the evolution of viruses. First, it permits the reconstruction of viable and fit genomes from unfit parents. This is a conservative evolutionary force that counteracts the deleterious effects of mutations. Second, it is a means of generating true genetic novelty, putting together either divergent forms of the same virus or genomic regions from two different parental viruses or even from viral and cellular nucleic acids. This is a radical evolutionary force that permits large evolutionary jumps. It is radical and risky in the sense of bringing together new combinations of genetic material, likely to be subjected to negative selection (see "Viral Quasispecies—Concept and Biological Implications" below).

Recombination plays an active role in present-day virus evolution and has most probably played an important role historically. Nucleotide sequence analyses have shown that several recent outbreaks of poliomyelitis have involved recombinants made up of poliovirus (PV) vaccine strains and other human enteroviruses, such as human enterovirus C. Recombination is also frequent among currently circulating enteroviruses (see, for example, reference 32; for a review, see reference 1). As a historical event, the alphavirus Western equine encephalitis virus probably originated by recombination between Eastern equine encephalitis-like and Sindbis-like viruses (for a review, see reference 69).

Recombination plays an important role in NoV evolution as well. The most frequent recombination site in NoV is the tightly conserved region surrounding the start codon of the capsid protein gene. Some polymerase types exist in combination with up to seven different capsid types, and some capsid types exist in combination with up to seven different polymerase types (56).

Most of the sequences that cannot be unambiguously assigned to established genotypes appear to contain either polymerase or capsid gene sequences from established genotypes. One could ask where all these genes come from: what is their reservoir? Food contaminated at the source with sewage containing multiple NoV genotypes provides an excellent way to fulfill one of the requirements for recombination, namely, simultaneous infection with at least two different viruses. This has been demonstrated to occur during outbreaks and infections due to ingestion of contaminated oysters (31, 41).

Segment Reassortment and Its Role in Virus Evolution

Segment reassortment occurs with RNA and DNA viruses whose genomes consist of two or more segments. On coinfection of a cell by two variants of a virus with a genome of n segments, a total of 2^n different reassortants can be produced. Fitness differences among the segment combinations generated will influence the types of reasssortants that participate in subsequent rounds of infection. The genome segments can be encapsidated either into a single particle (as, for example, in the orthomyxoviruses or arenaviruses) or into separate particles (as in multipartite plant viruses).

Reassortment as a major evolutionary force has been repeatedly documented with the type A human influenza viruses. Reassortment between human and animal influenza viruses has been the basis of antigenic shift, often associated with new influenza pandemics facilitated by a new antigenic composition of the reassortant virus (51, 70). Similarly, reassortment can occur in rotaviruses. Many human rotavirus strains have segments that are closely related to animal rotavirus genes. It is thought that new rotavirus strains emerge by cross-species transmission; during their adaptation to the new host, these emerging viruses reassort with established rotaviruses from the new host.

For recombination and reassortment to occur, a cell must be coinfected with at least two parental genomes; when the latter are genetically or phenotypically distinguishable, the occurrence of recombination or reassortment events can be detected. Persistence of a virus in a cell—or a slow kinetics of viral replication in relation to cell killing—may facilitate reinfections, unless some reinfection exclusion mechanism operates. Thus, the probability of recombination and reassortment is closely linked to the virus-host relationship. Recent evidence suggests that some cells in an infected host may be prone to coinfection, which would occur at higher frequency than expected on a random basis (8, 9). This may be relevant for food-borne viruses, since food-borne outbreaks most commonly involve infected food handlers who spread viruses by contaminating food during collection, processing, or serving. More troublesome in terms of the potential for recombination are the food-borne outbreaks in which contamination has occurred early in the food chain, during aquaculture (e.g, molluscan shellfish) or irrigation (e.g., raspberries).

DYNAMICS OF VIRUS EVOLUTION IN VIVO

The mechanisms of genetic variation of viruses described in the preceding sections are blind processes that generate repertoires of variants which are then subjected to selection and random (chance) expansions in each particular environment in which virus replication takes place. These events are guided by basic Darwinian principles that apply to all biological entities and that can be summarized as competition among variant forms, selection of the variants (or sets of related variants) best adapted to each environment, and random drift of genomes as a result of population bottlenecks of different intensities. Darwinian principles apply to both DNA and RNA viruses, irrespective of their mutation rates and population size. However, virus variations and adaptability have been most clearly documented with RNA viruses and some DNA viruses, in what is known as quasispecies dynamics.

This is where the application of these principles to food virology starts. In this section we discuss how some of these principles apply to food virology.

Virus Transmission via Food

Few if any viruses are transmitted solely via food. In general, virus transmission by food or water should be considered a chance event; it should be regarded as a special case of fecal-oral transmission, with the virus being present in food or the environment for short or long periods. The time interval is short when food is consumed immediately after being contaminated with a virus by a food handler. Longer intervals occur, for example, when a virus is present in the environment and is filtered from water by filter-feeding bivalve mollusks, which are subsequently harvested and consumed. Alternatively, long intervals can occur when fruit or vegetables are irrigated with sewage-contaminated water during growth and then frozen after being harvested. In this case, it may be consumed years later.

The length of the interval outside the host may not influence viral evolution, since replication does not occur. However, particles with increasing resistance to environmental stresses may be selected, leading to strains that are capable of surviving in the environment. For example, HAV, like other enteroviruses, shows a higher resistance to acidic pH than do other structurally and genetically related picornaviruses. In a study of inactivation of about 10^7 PFU, 10^4 PFU of HAV survived at pH 3.3 for 5 min (11). The process of strong selection of a minority of particles from a population can be regarded as a bottleneck event through which a few founder genomes constitute the seed for new population expansions. Bottleneck events have a number of implications for virus biology (15–17); for food-borne viruses, bottlenecks may be a means of restricting replication to subpopulations more likely to survive the serial food-host-food cycles involved in virus spread. Environmental factors may be mutagenic, in particular, UV light, certain chemicals, and products of oxidative stress within infected organisms. However, the impact of this mutagenesis on variation is generally regarded as minor compared to the effect of polymerase errors in RNA viruses. In

DNA viruses, UV light may introduce thymidine dimers, which are usually fatal for the virus.

Variation by mutation occurs in the host that sheds the virus. Selection at the genetic level occurs during the next replication round. Selection at the phenotype level can occur in between hosts, e.g., for viruses that are stable in the environment. A quasispecies is not only a complex mixture of genomes but also a mixture of pseudotypes, i.e., particles whose exterior protein does not match the enclosed genome.

Zoonotic Virus Transmission via Food

Transmission via food is an opportunity for an unlikely transmission event to occur relatively easily. Zoonoses with a fecal-oral route of infection are examples of this. For instance, hepatitis E virus types 3 and 4 are probably transmitted from pigs to humans (see also chapter 3) (42, 62, 63). There are indications that veterinarians who work with cattle are infected with cattle NoV; likewise, people who work with pigs could be infected with porcine NoV and sapoviruses, thus contributing to the gene pool of human viruses. To date, there is no evidence that this could happen for caliciviruses, but the existence of zoonotic transmission suggests that this should be further investigated (71).

The same applies to enteroviruses and rotaviruses, which are also present in cattle and pigs and are shed in their feces. Many rotavirus strains exist that have genome segments closely related to analogous segments of rotaviruses of other host species. Although there are many indications based on phylogenetic analysis that cross-species transmission events have occurred, there is no direct evidence in the literature; therefore, it is also speculative that the initial zoonotic transmission occurred as a food-borne event. The most compelling evidence that cross-species transmission has occurred was provided by analysis of the full complement of 11 segments of a human rotavirus isolate and a lapine isolate, which showed that all segments in the human isolate were most closely related to the lapine isolate (44).

An interesting example in which viral evolution appears to have contributed to its spread, and where introduction to the population may actually have been food borne, comes from the coronaviruses. Coronaviruses have long been known for their fecal-oral transmission route in different hosts. Closely related coronaviruses of humans and animal species exist, suggesting that cross-species transmission has occurred in the past, if not recently. Therefore, zoonotic (food-borne) transmission would not be surprising. During the search for the origin of SARS coronavirus, homologous antibodies have been detected in animal traders in southern China, in particular in traders of animals that were implicated in the transmission of SARS to humans. These animals were all traded for food. The most significant difference between animal and most human SARS coronavirus isolates in early 2003 was a deletion mutation in an area of the genome encoding accessory proteins. This deletion disrupts open reading frame 8ab, giving rise to two new expression products, 8a and 8b. Later it was found that the expression of the structural protein envelope (E) was downregulated by 8b but not by 8a or 8ab (39). These findings suggest that 8b may modulate viral replication and/or pathogenesis. Sequences of viral isolates involved in a separate cross-species transmission event in late 2003 did not have this deletion. The four human patients involved all had mild symptoms, suggesting that the deletion was an important factor in the severity of the first outbreak.

Cross-species transmission is an event with strong selective pressure at the level of virus entry. Since virus entry involves protein rather than nucleic acid, the relevance of genetic variation is limited if it does not result in phenotypic variation. A variant genome that contains a mutation enabling the virus to enter the new host may not allow the virus to enter the new host cell because it is packaged in a wild-type capsid. Therefore, if changes are required that play a role during entry and binding, these variants should already be amplified in the regular host to allow phenotypic changes to become apparent. A more likely scenario is that the initial cross-species transmission event is suboptimal and that subsequent adaptive evolution allows for more efficient replication and transmission in and among the new host species. After the initial transfer of SARS coronavirus to humans, the virus quickly adapted to its new host by mutation and positive selection. Most of the mutations identified were concentrated in a part of the spike protein gene that was later shown to be involved in receptor binding (72).

Food-Borne Seeding Event

Viruses with a low transmissibility are not likely to become epidemic. This can be expressed quantitatively by the basic reproductive ratio (Ro) or the average number of infected contacts from each infected host. Given the opportunity, an endemically circulating virus can become epidemic. A food-borne seeding event is such an opportunity. An example of this occurred in The Netherlands with a genotype IIc NoV (12). In January 2001, 231 persons from the staff of a department in The Netherlands became sick with diarrhea and vomiting after a buffet lunch which was prepared and served at a restaurant. An outbreak investigation showed an increasing risk of illness with the number of bread rolls eaten and found that the baker had vomited in the bakery sink on the day he prepared the rolls. The strain of virus detected had not been detected previously in outbreaks in The Netherlands. In the years before this outbreak, it was detected only once in a community case, approximately 22 months before the outbreak occurred in the same city. In the 6 months following this outbreak, however, this strain caused 13 outbreaks in The Netherlands and 21 outbreaks in England and Wales. Since July 2001, the same genotype has shown up only three times, once each in Norway, Sweden, and Germany, all in 2002. This suggests that this genotype has a limited survival in the environment and needed the seeding opportunity to become widespread; it was then able to cause a significant number of outbreaks. During the period when the virus was able to replicate more abundantly than when it was endemic, several outbreak strains were found that had mutations relative to the strain that caused the first outbreak. Although the mutations that we observed probably do not give a selective advantage, they illustrate the variability of epidemic strains. The calculated mutation rates are 4 and 10 changes per 1,000 nucleotides per year for England and Wales and for The Netherlands, respectively.

Contamination of Food

Food can be contaminated at the source or at any time during processing by a food handler. Contamination at the source occurs when irrigation water contains sewage or when filter-feeding bivalve mollusks grow in sewage-contaminated water. Transmission of viruses by food is often hard to prove. Sewage-contaminated food often contains many different viruses that are shed in feces. The host who eats the contaminated food may not necessarily become infected with the most abundant virus. Depending on host susceptibility, viruses present in smaller numbers may be preferentially selected by the host. In the past, epidemiologic evidence for certain food-borne outbreaks was discounted because there was no match between virus detected in the food and in the patients. This may be questionable in view of the above observations.

Viral Quasispecies—Concept and Biological Implications

Viral quasispecies are dynamic distributions of nonidentical but closely related viral genomes which are subjected to a continuous process of genetic variation, competition, and selection and which act as a unit of selection (Fig. 1) (reviewed in references 15 and 17). Mutant swarms are subjected to two classes of selection: (i) positive selection, which results in the dominance of the mutants (or mutant ensembles) that manifest an advantage in replication, and (ii) negative selection, which results in a reduction in the frequency (or even in the elimination) of the mutants (or mutant ensembles) that manifest a disadvantage in replication. The concept of quasispecies originated as a theory of self-organization and adaptability of early replicons and was intended to explain the origin of primitive life forms (23, 25). Early experimental

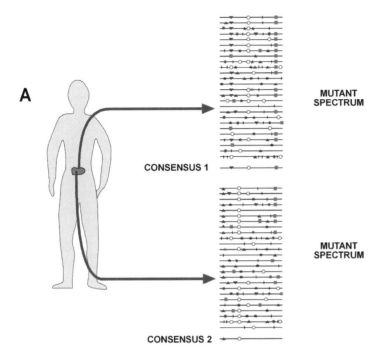

A

CONSENSUS 1

MUTANT
SPECTRUM

MUTANT
SPECTRUM

CONSENSUS 2

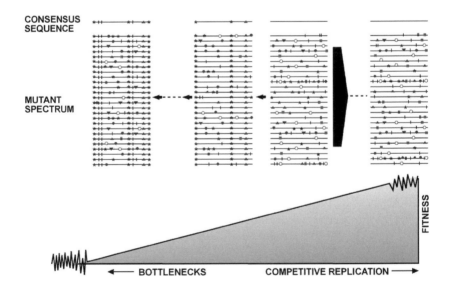

B

CONSENSUS
SEQUENCE

MUTANT
SPECTRUM

FITNESS

← BOTTLENECKS   COMPETITIVE REPLICATION →

studies with the *Escherichia coli* bacteriophage Qβ permitted the first calculation of a mutation rate for a specific nucleotide transition in an RNA virus (4) and the observation of a dynamic of mutant generation, competition, and selection that closely matched the predictions of quasispecies theory (20); a historical account can be found in reference 38. Subsequent investigations with many human, animal, and plant RNA viruses have documented quasispecies dynamics as a feature of RNA viruses and some DNA viruses. Quasispecies insights have contributed to the understanding of virus adaptability in relation to both pathogenesis and genetic and phenotypic diversification.

The main biological implications of viral quasispecies are as follows.

(i) Viral populations consist of a spectrum of complex and dynamic (continuously changing) mutants rather than defined genomic sequences (Fig. 1). The existence of swarms of related genomes increases their adaptive potential because individual viral variants or sets of minority variants may manifest a growth advantage in the face of an environmental change (presence of antibodies, cytotoxic T lymphocytes, drugs, etc.). It has been shown that in some cases the complexity of the mutant spectrum can affect the outcome of an infection (26) or the response to therapy (19, 27, 52, 53). A PV mutant expressing a polymerase with enhanced template-copying fidelity relative to the wild type generated a narrower mutant spectrum and, unlike wild-type PV, did not cause neuropathology in susceptible mice (55, 64).

(ii) The spectrum of mutants does not behave as a conglomerate of independent genomes; instead, positive or negative intrapopulation interactions are frequently established that give rise to a collective behavior that cannot be predicted from the behavior of the individual components. Positive interactions permit the survival of some variants by complementation (47). A PV mutant which could not reach the brains of mice when inoculated alone could do so when inoculated together with a neurovirulent PV quasispecies (64). Suppression of infectivity can be mediated by defective, *trans*-acting products expressed by mutants present in the spectrum of mutants (10, 34, 35, 54). Additional cases of suppression of specific variants have been documented in

---

Figure 1 Quasispecies are dynamic distributions of mutant and recombinant genomes characterized by a mutant spectrum and a consensus sequence. In this scheme, genomes are depicted as horizontal lines and mutations are depicted as symbols on the lines. (A) Multiple quasispecies can coexist in an infected organism, even within an organ. (B) Large population passages (large black arrow) lead to competitive optimization of the quasispecies and to fitness gain in the environment being considered (triangle at the bottom). In contrast, bottleneck events (small arrows) lead to random accumulation of mutations (compare consensus sequences) and a decrease in fitness. At high and low fitness values, fluctuations of fitness values may occur. (Adapted from reference 14 with permission.)

vivo and in model studies of cell cultures. These observations conform to the concept of an ensemble of genomes acting as a unit of selection, as predicted by quasispecies theory.

(iii) A consequence of quasispecies dynamics is the presence of minority genomes in viral quasispecies that reflect the genomes that were dominant at an earlier stage of the same evolutionary lineage. Memory genomes have been characterized in cell culture and in vivo with foot-and-mouth disease virus FMDV and human immunodeficiency virus type 1 HIV-1 (5, 6, 57). The frequency of memory genomes was 10^{-1} to 10^{-2}, a value generally higher than the level that could be attributed merely to mutation pressure inherent to the replication machinery. Memory levels were fitness dependent, and memory was erased by intervening bottleneck events. Quasispecies memory can provide viral populations with a replicative advantage when they are confronted with a selective constraint that had already been experienced by the same viral lineage at an earlier stage of its evolution (15, 57). Memory illustrates, again, a feature of viral quasispecies that depends on the mutant distribution as an ensemble.

(iv) Quasispecies theory has impacted viral disease prevention and treatment on several fronts. First, mutation rates and frequencies readily explain the selection of vaccine escape or inhibitor escape mutants, unless some conditions on vaccination and antiviral treatments are met. Vaccines must evoke a broad immune response directed to multiple B-cell and T-cell epitopes represented in the authentic infectious virus to be controlled. For basic statistical reasons, a highly variable virus will probably escape an immune response directed to one (or very few) epitopes, as documented by the frequent isolation of monoclonal antibody (MAb)-escape mutants. Likewise, the use of combination therapy with three or more antiviral inhibitors directed to different viral targets is necessary to prevent or delay the selection of inhibitor-resistant viral variants (specific examples and further quantitative justification are given in references 13 and 17; the successive versions of reference 13 provide an account of the development of this problem over three decades). Second, a prediction of quasispecies theory is that an increase in the mutation rate of a virus should lead to loss of its genetic information and therefore to its extinction (24, 25). This has led to the development of a new antiviral strategy termed lethal mutagenesis, which exploits the administration of mutagenic base (or nucleoside) analogues as antiviral agents. The antiviral effect of mutagenic agents, associated with their mutagenic activity, has been demonstrated with several virus-host systems both in cell culture and in vivo (for reviews, see references 2, 14, and 16).

The current picture of the mechanism of lethal mutagenesis is that as the mutation rate during viral replication is increased, lethal mutations and nonlethal but deleterious mutations increase in frequency. Mutations can be deleterious for replication of the genome harboring them or for replication of peer genomes (interference). Accumulation of lethal and interfering mutations is reflected in decreases in specific infectivity and an invariant consensus sequence. The latter observation is expected from the fact that no specific genomes are selected when lethal mutagenesis is effective and that no extinction escape mutants become dominant. Defective genomes promote extinction, probably through *trans*-active defective gene products that interfere with one or several steps in the virus life cycle (34, 35, 54). Lethal mutagenesis constitutes an active area of research. In 2005 the first clinical trial involving administration of a mutagenic nucleoside analogue to AIDS patients (who had failed to respond to highly active antiretroviral therapy) was initiated (37), and this trial was scheduled to enter phase II in 2007. A critical issue is whether selection of viral polymerase mutants with decreased sensitivity to nucleoside analogues, recently documented with PV and foot-and-mouth disease virus (55, 60, 64), can jeopardize lethal-mutagenesis approaches. This is an important question that is being actively investigated in several laboratories.

Thus, high mutation rates and quasispecies dynamics have helped explain the adaptive potential of many pathogenic viruses and have set the stage for the application of modified and new antiviral designs. The adaptative potential of RNA viruses has a number of implications for food-borne and waterborne viral illness and its control.

ADAPTABILITY THROUGH SELECTION OF VIRAL SUBPOPULATIONS AND VIRUS PERSISTENCE IN THE ENVIRONMENT

Viruses that cause disease must have been selected to persist in our biosphere. Persistence of viral pathogens has at least two meanings. One is long-term survival in nature, mediated by a balance between replication of the viruses and multiplication of their host organisms. The other, the most usual meaning of persistence, is the survival of a virus within an individual host organism for prolonged periods. In this case, survival is ensured because damage to the host organism is either delayed, limited, or nonexistent.

Balance of a viral pathogen with multiple host individuals for long-term survival in nature often involves cycles of intrahost replication and interhost transmission. The essential transmission step can occur through direct contact between infected and susceptible individuals or through environmental contamination as the main mechanisms. Virus stability in the environment can affect viral virulence. For respiratory tract pathogens, virulence was correlated with durability in the external environment, and the correlation was interpreted as high environmental durability reducing the dependence of transmission on host mobility (67). Durability of viral pathogens in the environment, including food and water, is favored by the amount of virus present, which in turn is dependent on the replication capacity in the host organism that is the source of the environmental contamination. For a given initial amount of virus, durability also depends on particle stability (see chapter 8), which is in turn dependent on interactions among capsid (or surface protein) residues. Interactions that stabilize viral particles can be subjected to variation, competition, and selection; therefore, a high or low stability may be selected, depending on the time a virus spends as free particles and inside the cell. It may be relevant to explore connections between capsid structure (based on phylogenetic position) and stability in the environment or resistance to inactivating processes such as pressure-temperature cycles (40).

An RNA virus that contaminates food or water is expected to contain a spectrum of mutant and recombinant genomes on which selection can act at different levels. First, subpopulations that display higher particle stability will survive for longer periods and have a higher probability of infecting a susceptible host. Second, of all subpopulations present, some may be selectively transmitted because they are endowed with a higher potential for replication in a recipient host, as recently suggested by transmission studies with blood contaminated with hepatitis C virus (49).

The variability of RNA viruses has several implications regarding the control and prevention of food-borne and waterborne viral infections. One is that an identical nucleotide sequence is not to be expected when comparing a putative source virus (for example, contaminated food or water) and the virus found in the infected recipients. High mutation rates make the complete conservation of sequences very unlikely as soon as virus replication occurs, in this case in the infected individuals. Another consequence is that identical susceptibility of food-borne virus strains to a disinfection procedure cannot be assumed. Variants of the same virus, even if differing only minimally in their genomes, may display different sensitivities to a disinfection procedure.

CONCLUSION

Historically, the continuous cycles of genetic variation, generation of quasispecies swarms, selection of virus subpopulations through bottlenecks of various intensities, and environment-dependent, differential survival have contributed to the development of constellations of phenotypic traits (13–19). Among such traits, particle stability in food or water, together with resistance to the gastrointestinal environment, has given rise to the current food-borne viral pathogens. It must be considered, however, that viruses—in particular RNA viruses—are undergoing continuous evolution and that new food-borne variants may be selected in an increasingly complex and dynamic world. Thus, evolution has been of great help in tracing the origin of disease outbreaks through genetic analyses, but its activity on rapidly replicating RNA viruses may be contributing to the emergence of viral pathogens or to pathogens with new transmission routes, selected among myriads of variants.

REFERENCES

1. **Agol, V. I.** 2006. Molecular mechanisms of poliovirus variation and evolution. *Curr. Top. Microbiol. Immunol.* **299:**211–259.

2. **Anderson, J. P., R. Daifuku, and L. A. Loeb.** 2004. Viral error catastrophe by mutagenic nucleosides. *Annu. Rev. Microbiol.* **58:**183–205.

3. **Arnold, J. J., M. Vignuzzi, J. K. Stone, R. Andino, and C. E. Cameron.** 2005. Remote site control of an active site fidelity checkpoint in a viral RNA-dependent RNA polymerase. *J. Biol. Chem.* **280:**25706–25716.

4. **Batschelet, E., E. Domingo, and C. Weissmann.** 1976. The proportion of revertant and mutant phage in a growing population, as a function of mutation and growth rate. *Gene* **1:**27–32.

5. **Briones, C., A. de Vicente, C. Molina-Paris, and E. Domingo.** 2006. Minority memory genomes can influence the evolution of HIV-1 quasispecies in vivo. *Gene* **384:**129–138.

6. **Briones, C., E. Domingo, and C. Molina-Paris.** 2003. Memory in retroviral quasispecies: experimental evidence and theoretical model for human immunodeficiency virus. *J. Mol. Biol.* **331:**213–229.

7. **Chetverin, A. B., D. S. Kopein, H. V. Chetverina, A. A. Demidenko, and V. I. Ugarov.** 2005. Viral RNA-directed RNA polymerases use diverse mechanisms to promote recombination between RNA molecules. *J. Biol. Chem.* **280:**8748–8755.

8. **Chohan, B., L. Lavreys, S. M. Rainwater, and J. Overbaugh.** 2005. Evidence for frequent reinfection with human immunodeficiency virus type 1 of a different subtype. *J. Virol.* **79:**10701–10708.

9. **Cicin-Sain, L., J. Podlech, M. Messerle, M. J. Reddehase, and U. H. Koszinowski.** 2005. Frequent coinfection of cells explains functional in vivo complementation between cytomegalovirus variants in the multiply infected host. *J. Virol.* **79:**9492–502.

10. **Crowder, S., and K. Kirkegaard.** 2005. Trans-dominant inhibition of RNA viral replication can slow growth of drug-resistant viruses. *Nat. Genet.* **37:**701–709.

11. **Deboosere, N., O. Legeay, Y. Caudrelier, and M. Lange.** 2004. Modelling effect of physical and chemical parameters on heat inactivation kinetics of hepatitis A virus in a fruit model system. *Int. J. Food Microbiol.* **93:**73–85.

12. **de Wit, M. A., M. A. Widdowson, H. Vennema, E. de Bruin, T. Fernandes, and M. Koopmans.** 2007. Large outbreak of norovirus: the baker who should have known better. *J. Infect.* **55:**188–193.

13. **Domingo, E.** 2003. Quasispecies and the development of new antiviral strategies. *Prog. Drug Res.* **60:**133–158.

14. **Domingo, E. (ed.).** 2005. Virus entry into error catastrophe as a new antiviral strategy. *Virus Res.* **107**:115–228.

15. **Domingo, E., V. Martin, C. Perales, A. Grande-Perez, J. Garcia-Arriaza, and A. Arias.** 2006. Viruses as quasispecies: biological implications. *Curr. Top. Microbiol. Immunol.* **299**:51–82.

16. **Domingo, E.** 2007. Virus evolution, p. 389–421. *In* D. M. Knipe, P. M. Howley, D. E. Griffin, R. A. Lamb, M. A. Martin, B. Roizman, and S. E. Straus (ed.), *Fields Virology*, 5th ed. Lippincott Williams & Wilkins, Philadelphia, PA.

17. **Domingo, E., C. Biebricher, M. Eigen, and J. J. Holland.** 2001. *Quasispecies and RNA Virus Evolution: Principles and Consequences.* Landes Bioscience, Austin, TX.

18. **Domingo, E., A. Brun, J. I. Núñez, J. Cristina, C. Briones, and C. Escarmís.** 2006. Genomics of viruses, p. 369–388. *In* J. Hacker and U. Dobrindt (ed.), *Pathogenomics: Genome Analysis of Pathogenic Microbes.* Wiley-VCH Verlag GmbH & Co. KGaA, Weinheim, Germany.

19. **Domingo, E., and J. Gomez.** 2007. Quasispecies and its impact on viral hepatitis. *Virus Res.* **127**:131–150.

20. **Domingo, E., D. Sabo, T. Taniguchi, and C. Weissmann.** 1978. Nucleotide sequence heterogeneity of an RNA phage population. *Cell* **13**:735–744.

21. **Drake, J. W., and J. J. Holland.** 1999. Mutation rates among RNA viruses. *Proc. Natl. Acad. Sci. USA* **96**:13910–13913.

22. **Earl, D. J., and M. W. Deem.** 2004. Evolvability is a selectable trait. *Proc. Natl. Acad. Sci. USA* **101**:11531–11536.

23. **Eigen, M.** 1971. Self-organization of matter and the evolution of biological macromolecules. *Naturwissenschaften* **58**:465–523.

24. **Eigen, M.** 2002. Error catastrophe and antiviral strategy. *Proc. Natl. Acad. Sci. USA* **99**:13374–13376.

25. **Eigen, M., and P. Schuster.** 1979. *The Hypercycle. A Principle of Natural Self-Organization.* Springer-Verlag KG, Berlin, Germany.

26. **Farci, P., A. Shimoda, A. Coiana, G. Diaz, G. Peddis, J. C. Melpolder, A. Strazzera, D. Y. Chien, S. J. Munoz, A. Balestrieri, R. H. Purcell, and H. J. Alter.** 2000. The outcome of acute hepatitis C predicted by the evolution of the viral quasispecies. *Science* **288**:339–344.

27. **Farci, P., R. Strazzera, H. J. Alter, S. Farci, D. Degioannis, A. Coiana, G. Peddis, F. Usai, G. Serra, L. Chessa, G. Diaz, A. Balestrieri, and R. H. Purcell.** 2002. Early changes in hepatitis C viral quasispecies during interferon therapy predict the therapeutic outcome. *Proc. Natl. Acad. Sci. USA* **99**:3081–3086.

28. **Ferrer-Orta, C., A. Arias, C. Escarmis, and N. Verdaguer.** 2006. A comparison of viral RNA-dependent RNA polymerases. *Curr. Opin. Struct. Biol.* **16**:27–34.

29. **Forterre, P.** 2006. The origin of viruses and their possible roles in major evolutionary transitions. *Virus Res.* **117**:5–16.

30. **Friedberg, E. C., G. C. Walker, W. Siede, R. D. Wood, R. A. Schultz, and T. Ellenberger.** 2006. *DNA Repair and Mutagenesis.* American Society for Microbiology, Washington, DC.

31. **Gallimore, C. I., J. S. Cheesbrough, K. Lamden, C. Bingham, and J. J. Gray.** 2005. Multiple norovirus genotypes characterised from an oyster-associated outbreak of gastroenteritis. *Int. J. Food Microbiol.* **103**:323–330.

32. **Gavrilin, G. V., E. A. Cherkasova, G. Y. Lipskaya, O. M. Kew, and V. I. Agol.** 2000. Evolution of circulating wild poliovirus and of vaccine-derived poliovirus in an immunodeficient patient: a unifying model. *J. Virol.* **74:**7381–7390.

33. **Gmyl, A. P., S. A. Korshenko, E. V. Belousov, E. V. Khitrina, and V. I. Agol.** 2003. Nonreplicative homologous RNA recombination: promiscuous joining of RNA pieces? *RNA* **9:**1221–1231.

34. **González-López, C., A. Arias, N. Pariente, G. Gómez-Mariano, and E. Domingo.** 2004. Preextinction viral RNA can interfere with infectivity. *J. Virol.* **78:**3319–3324.

35. **Grande-Pérez, A., E. Lazaro, P. Lowenstein, E. Domingo, and S. C. Manrubia.** 2005. Suppression of viral infectivity through lethal defection. *Proc. Natl. Acad. Sci. USA* **102:** 4448–4452.

36. **Green, K. Y., R. M. Chanock, A. Z. Kapikian.** 2001. Human caliciviruses, p. 841–874. *In* D. M. Knipe and P. M. Howley (ed.), *Fields Virology,* 4th ed. Lippincott Williams & Wilkins, Philadelphia, PA.

37. **Harris, K. S., W. Brabant, S. Styrchak, A. Gall, and R. Daifuku.** 2005. KP-1212/1461, a nucleoside designed for the treatment of HIV by viral mutagenesis. *Antiviral Res.* **67:**1–9.

38. **Holland, J.** 2006. Transitions in understanding of RNA viruses: an historical perspective. *Curr. Top. Microbiol. Immunol.* **299:**371–401.

39. **Keng, C. T., Y. W. Choi, M. R. Welkers, D. Z. Chan, S. Shen, S. Gee Lim, S. Hong, and Y. J. Tan.** 2006. The human severe acute respiratory syndrome coronavirus (SARS-CoV) 8b protein is distinct from its counterpart in animal SARS-CoV and down-regulates the expression of the envelope protein in infected cells. *Virology* **354:**132–142.

40. **Kingsley, D. H., D. Guan, D. G. Hoover, and H. Chen.** 2006. Inactivation of hepatitis A virus by high-pressure processing: the role of temperature and pressure oscillation. *J. Food Prot.* **69:**2454–2459.

41. **Le Guyader, F. S., F. Bon, D. DeMedici, S. Parnaudeau, A. Bertone, S. Crudeli, A. Doyle, M. Zidane, E. Suffredini, E. Kohli, F. Maddalo, M. Monini, A. Gallay, M. Pommepuy, P. Pothier, and F. M. Ruggeri.** 2006. Detection of multiple noroviruses associated with an international gastroenteritis outbreak linked to oyster consumption. *J. Clin. Microbiol.* **44:**3878–3882.

42. **Li, T. C., K. Chijiwa, N. Sera, T. Ishibashi, Y. Etoh, Y. Shinohara, Y. Kurata, M. Ishida, S. Sakamoto, N. Takeda, and T. Miyamura.** 2005. Hepatitis E virus transmission from wild boar meat. *Emerg. Infect. Dis.* **11:**1958–1960.

43. **Martin, D. P., C. Williamson, and D. Posada.** 2005. RDP2: recombination detection and analysis from sequence alignments. *Bioinformatics* **21:**260–262.

44. **Matthijnssens, J., M. Rahman, V. Martella, Y. Xuelei, S. De Vos, K. De Leener, M. Ciarlet, C. Buonavoglia, and M. Van Ranst.** 2006. Full genomic analysis of human rotavirus strain B4106 and lapine rotavirus strain 30/96 provides evidence for interspecies transmission. *J. Virol.* **80:**3801–3810.

45. **Menéndez-Arias, L.** 2002. Molecular basis of fidelity of DNA synthesis and nucleotide specificity of retroviral reverse transcriptases. *Prog. Nucleic Acid Res. Mol. Biol.* **71:**91–147.

46. **Minskaia, E., T. Hertzig, A. E. Gorbalenya, V. Campanacci, C. Cambillau, B. Canard, and J. Ziebuhr.** 2006. Discovery of an RNA virus $3' \rightarrow 5'$ exoribonuclease that is critically involved in coronavirus RNA synthesis. *Proc. Natl. Acad. Sci. USA* **103:**5108–5113.

47. **Moreno, I. M., J. M. Malpica, E. Rodriguez-Cerezo, and F. Garcia-Arenal.** 1997. A mutation in tomato aspermy cucumovirus that abolishes cell-to-cell movement is maintained to high levels in the viral RNA population by complementation. *J. Virol.* **71:**9157–9162.

48. **Nagy, P. D., and A. E. Simon.** 1997. New insights into the mechanisms of RNA recombination. *Virology* **235:**1–9.

49. **Nainan, O. V., L. Lu, F. X. Gao, E. Meeks, B. H. Robertson, and H. S. Margolis.** 2006. Selective transmission of hepatitis C virus genotypes and quasispecies in humans and experimentally infected chimpanzees. *J. Gen. Virol.* **87:**83–91.

50. **Page, R. D. M., and E. C. Holmes.** 1998. *Molecular Evolution. A Phylogenetic Approach.* Blackwell Science Ltd., Oxford, United Kingdom.

51. **Parrish, C. R., and Y. Kawaoka.** 2005. The origins of new pandemic viruses: the acquisition of new host ranges by canine parvovirus and influenza A viruses. *Annu. Rev. Microbiol.* **59:**553–586.

52. **Pawlotsky, J. M.** 2000. Hepatitis C virus resistance to antiviral therapy. *Hepatology* **32:**889–896.

53. **Pawlotsky, J. M.** 2006. Hepatitis C virus population dynamics during infection. *Curr. Top. Microbiol. Immunol.* **299:**261–284.

54. **Perales, C., R. Mateo, M. G. Mateu, and E. Domingo.** 2007. Insights into RNA virus mutant spectrum and lethal mutagenesis events: replicative interference and complementation by multiple point mutants. *J. Mol. Biol.* **369:**985–1000.

55. **Pfeiffer, J. K., and K. Kirkegaard.** 2005. Increased fidelity reduces poliovirus fitness under selective pressure in mice. *PLoS Pathogens* **1:**102–110.

56. **Phan, T. G., K. Kaneshi, Y. Ueda, S. Nakaya, S. Nishimura, A. Yamamoto, K. Sugita, S. Takanashi, S. Okitsu, and H. Ushijima.** 2007. Genetic heterogeneity, evolution, and recombination in noroviruses. *J. Med. Virol.* **79:**1388–1400.

57. **Ruiz-Jarabo, C. M., A. Arias, E. Baranowski, C. Escarmís, and E. Domingo.** 2000. Memory in viral quasispecies. *J. Virol.* **74:**3543–3547.

58. **Salemi, M., and A. M. Vandamme (ed.).** 2004. *The Phylogeny Handbook. A Practical Approach to DNA and Protein Phylogeny.* Cambridge University Press, Cambridge, United Kingdom.

59. **Sanchez, G., A. Bosch, G. Gomez-Mariano, E. Domingo, and R. M. Pinto.** 2003. Evidence for quasispecies distributions in the human hepatitis A virus genome. *Virology* **315:**34–42.

60. **Sierra, M., A. Airaksinen, C. González-López, R. Agudo, A. Arias, and E. Domingo.** 2007. Foot-and-mouth disease virus mutant with decreased sensitivity to ribavirin: implications for error catastrophe. *J. Virol.* **81:**2012–2024.

61. **Steinhauer, D. A., E. Domingo, and J. J. Holland.** 1992. Lack of evidence for proofreading mechanisms associated with an RNA virus polymerase. *Gene* **122:**281–288.

62. **Takahashi, K., N. Kitajima, N. Abe, and S. Mishiro.** 2004. Complete or near-complete nucleotide sequences of hepatitis E virus genome recovered from a wild boar, a deer, and four patients who ate the deer. *Virology* **330:**501–505.

63. **Tei, S., N. Kitajima, K. Takahashi, and S. Mishiro.** 2003. Zoonotic transmission of hepatitis E virus from deer to human beings. *Lancet* **362:**371–373.

64. **Vignuzzi, M., J. K. Stone, J. J. Arnold, C. E. Cameron, and R. Andino.** 2006. Quasispecies diversity determines pathogenesis through cooperative interactions in a viral population. *Nature* **439:**344–348.

65. **Villarreal, L. P.** 2005. *Viruses and the Evolution of Life.* ASM Press, Washington, DC.

66. **Vinje, J., J. Green, D. C. Lewis, C. I. Gallimore, D. W. Brown, and M. P. Koopmans.** 2000. Genetic polymorphism across regions of the three open reading frames of "Norwalk-like viruses." *Arch. Virol.* **145:**223–241.

67. **Walther, B. A., and P. W. Ewald.** 2004. Pathogen survival in the external environment and the evolution of virulence. *Biol. Rev. Camb. Philos. Soc.* **79:**849–869.

68. **Wang, Q. H., M. G. Han, S. Cheetham, M. Souza, J. A. Funk, and L. J. Saif.** 2005. Porcine noroviruses related to human noroviruses. *Emerg. Infect. Dis.* **11:**1874–1881.

69. **Weaver, S. C.** 2006. Evolutionary influences in arboviral disease. *Curr. Top. Microbiol. Immunol.* **299:**285–314.

70. **Webster, R. G.** 1999. Antigenic variation in influenza viruses, p. 377–390. *In* E. Domingo, R. G. Webster, and J. J. Holland (ed.), *Origin and Evolution of Viruses.* Academic Press, San Diego, CA.

71. **Widdowson, M. A., B. Rockx, R. Schepp, W. H. van der Poel, J. Vinjé, Y. T. van Duynhoven, and M. P. Koopmans.** 2005. Detection of serum antibodies to bovine norovirus in veterinarians and the general population in The Netherlands. *J. Med. Virol.* **76:**119–128.

72. **Zhang, C. Y., J. F. Wei, and S. H. He.** 2006. Adaptive evolution of the spike gene of SARS coronavirus: changes in positively selected sites in different epidemic groups. *BMC Microbiol.* **6:**88.

73. **Zheng, D. P., T. Ando, R. L. Fankhauser, R. S. Beard, R. I. Glass, and S. S. Monroe.** 2006. Norovirus classification and proposed strain nomenclature. *Virology* **346:**312–323.

Food-Borne Viruses: Progress and Challenges
Edited by Marion P. G. Koopmans, Dean O. Cliver, and Albert Bosch
© 2008 ASM Press, Washington, DC

Rethinking Virus Detection in Food

Rosa M. Pintó and Albert Bosch

Well over 100 different enteric viruses are liable to be found as food contaminants; however, with few exceptions most well-characterized food-borne viral outbreaks are restricted to calicivirus, essentially norovirus (NoV) and hepatitis A virus (HAV) (Table 1) (56), which in consequence are the main targets for virus detection in food.

NoV infections are very common (53, 56) and are likely to become more so, with new emerging strains being described with increasing frequency (see chapter 6) (52). In contrast, HAV, with only a single serotype so far described and for which a reliable vaccine is available, appears a potentially eradicable contaminant. However, in spite of its declining incidence in developed countries, HAV still causes around half of the acute hepatitis infections globally and remains more than a troublesome nuisance in developed communities. Systematic immunological surveys indicate that over 90% of the low-income population in developing countries bears serological evidence of infection by HAV before adulthood (2, 28, 31). Moreover, HAV causes occasional severe cases of fulminant hepatitis, with fatal outcomes in otherwise healthy adults (55).

Although many NoV and HAV infections may be transmitted from person to person, food-borne transmission remains an underrated route. Even when a viral outbreak is epidemiologically linked to food consumption, the lack of sensitive and reliable detection methods complicates confirmation based on laboratory isolation of the virus from the food product.

Cell culture propagation of wild-type strains of HAV is a complex and tedious task which requires virus adaptation before it can be grown effectively; even in this case, the virus usually establishes persistent infections

ROSA M. PINTÓ AND ALBERT BOSCH, Enteric Virus Laboratory, Department of Microbiology, University of Barcelona, 08028 Barcelona, Spain.

Table 1 Examples of documented NoV and HAV food-borne outbreaks.

Agent	Yr	Foodstuff	No. of cases	Location	Food origin	Reference
NoV	1988	Raspberries	108	Helsinki (Finland)	Unknown	69
	1993	Oysters	190	United States	Louisiana (United States)	49
	1996	Oysters	75	United States	Louisiana (United States)	7
		Oysters	153	United States	Louisiana (United States)	7
	1997	Raspberries	200	Quebec (Canada)	Unknown	32
HAV	1979	Mussels	41	Leeds (United Kingdom)	Ireland	11
	1988	Clams	300,000	Shanghai (China)	China	34
		Lettuce	202	Kentucky (United States)	Kentucky (United States)	73
	1997	Strawberries	153	Michigan (United States)	Mexico (processed in the United States)	15
	1998	Green onions	43	Ohio (United States)	Mexico and California (United States)	23
	1999	Clams	184	Valencia (Spain)	Peru	75
	2002	Blueberries	39	Auckland (New Zealand)	New Zealand	13
	2003	Green onions	600	Pennsylvania (United States)	Mexico	84
	2005	Oysters	39	United States	Mississippi (United States)	76

resulting in low virus yields (22, 29). Recently, the use of a cell line that allows the growth of a wild-type HAV isolate from stool samples has been reported (50), although its validity for broad-spectrum isolation of HAV is not yet demonstrated. For NoV, the situation is even worse; in spite of numerous attempts (3, 25, 86), NoV propagation has been almost impossible until very recently through the use of an organoid model of human small intestinal epithelium that provides promising results for infectious NoV detection (77). In conclusion, until issues regarding assay complexity, cost-effectiveness, and validity for the detection of a broad spectrum of isolates have been addressed, infectivity is not nowadays a useful method for NoV and HAV detection.

Regarding immunological techniques used in clinical diagnosis, for HAV these are aimed at the detection of anti-HAV antibodies generated by virus infection, mainly immunoglobulin M and immunoglobulin G antigen (58, 59), while antigen detection is uncommon. Systems for NoV detection in fecal specimens do exist (5); however, even when kits for antigen detection are available, their sensitivity is not high enough for them to be employed in scenarios of low virus concentration, such as in most foodstuffs.

All the above-mentioned caveats make nucleic acid amplification techniques the obvious choice for the development of highly sensitive procedures for virus detection in food.

MOLECULAR APPROACHES TO VIRUS DETECTION IN FOOD—PERSPECTIVES AND LIMITATIONS

Nucleic acid amplification techniques are currently the most widely used methods for detection of viruses in food. Although nucleic acid sequence-based amplification and loop-mediated isothermal amplification techniques have been reported to be highly sensitive and specific, respectively (30, 40, 41, 61), reverse transcription-PCR (RT-PCR) remains the current "gold standard" for virus detection in food. A dramatic improvement comes from the emergence of real-time RT-PCR, which makes use of fluorescent probes and enables not only qualitative determination but also quantitative diagnostic assays. Although the generic determination of pathogens is the essence of diagnostic practices, the possibility of quantitative virus detection represents a major advance in routine monitoring in virology.

Both HAV and NoV are RNA viruses and thus depend on error-prone polymerases, which lead to complex mutant genome populations or quasispecies (24). The selection of a highly conserved primer-probe set should be the first step in the design of a real-time RT-PCR assay. This goal, however, while attainable for HAV, is not so easily reached for NoV. RNA regions containing complex multidomain structures, mostly involved in translation (26) and/or replication (67) functions, must be highly conserved.

Like all picornaviruses, HAV, the prototype of the *Hepatovirus* genus in this family, contains highly conserved RNA regions at each end of its genome: the 5′ noncoding region (5′NCR) and the 3′NCR. Phylogenetic analyses of the 5′NCR of picornavirus reveal extensive structure-conserving substitutions within predicted stems, a high degree of sequence conservation in predicted loops, and clustering of regions with sequence divergence in spacer regions between domains (70). Picornavirus translation is cap independent, and the recruitment of the translation machinery takes place at the internal ribosome entry site contained in the 5′NCR (Fig. 1); hence, this latter region is particularly suitable for use as a target for amplification in diagnostic assays (19, 21, 36, 42). To ascertain the robustness of the primer-probe target sequences, it is advisable to perform quasispecies analyses prior to the development of quantitative procedures (19).

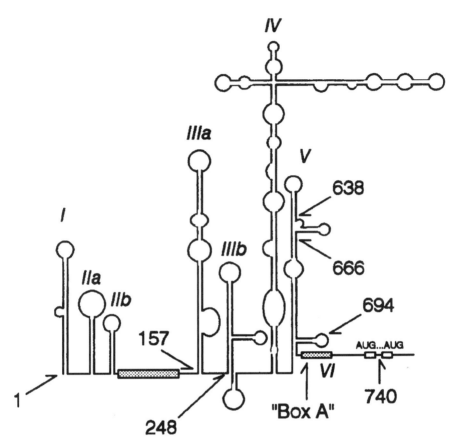

Figure 1 Secondary structure of the 5'NCR RNA of HAV strain HM175. (Reprinted from reference 12 with permission.)

The absence of a 5'NCR, and in particular of an IRES, in the calicivirus genomic and subgenomic RNAs makes the approach described for HAV unfeasible for use with NoV. Regions with structured elements such as the 3'NCR, or even the ribosome frameshifting between the capsid and the terminal protein regions, could be appropriate alternatives provided that their lengths were sufficient to include primers and a probe located at the proper distance. However, recent evidence points to a mechanism of de novo initiation of translation for the synthesis of the VP2 protein of the feline calicivirus rather than to an actual frameshifting (54), which would result in a short structured element similar to those likely to occur at the 5' ends of genomic and subgenomic RNAs. Regarding the 3'NCR, the predicted length ranges from 42 to 82 in human caliciviruses (33), which reduces its usefulness as an appropriate amplification target. As a consequence, multiplex approaches involving several primer sets and probes are required to address the diversity of these viruses, and no single assay stands out as the best by all criteria (5, 82). When using coding regions for amplification, an alternative to the multiplex systems is the design of primer-probe sets that include codons for methionine and tryptophan (single-codon amino acid families), since any mutation at these sites would be nonsynonymous and hence less prone to take place. Recent molecular data have demonstrated the presence of relatively highly conserved regions at the 3' end of open reading frame 1 (ORF1) (polymerase) and the 5' end of ORF2 (capsid) (Fig. 2) (16, 44, 47, 83), whereby the ORF1-ORF2 junction is proposed as a target for real-time amplification (38, 44–46, 51, 60, 80).

The aforementioned approaches to the development of broadly reactive detection methods do not provide the ability to reach another ultimate goal in diagnostics, which is the possibility to type the isolates. This objective is particularly relevant for the highly variable NoV, and most reported typing systems are based on sequence analysis of polymerase and capsid regions after standard RT-PCR assays (81–83). For HAV, the most widely used

Figure 2 Analysis of the variability of full-length NoV genomes of genogroup I (A) and genogroup II (B) strains. The locations of the ORFs are noted. Thick lines depict the most highly conserved regions. The average similarity score within each genogroup is represented as a dotted line. (Reprinted from reference 44 with permission.)

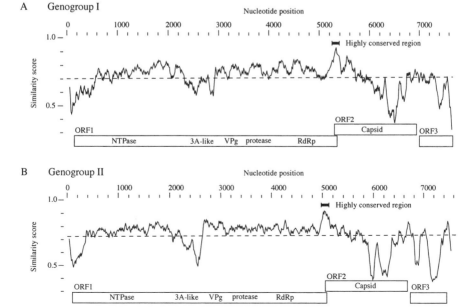

region is the amino-terminal 2A region, which is included in a larger fragment corresponding to the VP1-VP2A junction region (72). However, other data suggest that analysis of the entire VP1 region would lead to improved type discrimination (20, 78).

TARGETED FOOD MATRICES: WHICH MATRICES MAY BE REALISTICALLY ASSAYED FOR CONTAMINANT VIRUSES?

A wide variety of foodstuffs may become contaminated by viruses during the farm-to-fork chain, during either the preharvest or postharvest stages. Among the foods at risk of preharvest contamination are bivalve shellfish, fresh produce, and water. Postharvest contamination is most likely to result from poor hygiene practices during food handling, and hence the foods most at risk are uncooked or lightly cooked foods.

In contrast to other pathogen or chemical contaminants, foods are rarely tested for viral contamination, and such testing, when performed, is usually limited to shellfish. The main reason for this lack of virological analysis is the absence of legal regulations specifying such control, which is a direct consequence of several factors such as the important technical limitations that have hampered the development of reference methods for virus detection in food and, last but not least, the increased cost of the food product. However, the benefit of reducing health problems through consumption of safe food products calls for efforts in clearly defining which food matrices should be tested for viruses. In this context, foods of primary importance are those likely to be contaminated at the preharvest stage such as bivalve mollusks, particularly oysters, clams, and mussels; salad crops such as lettuce, green onions, and other greens; and soft fruits, such as raspberries and strawberries. All these types of food have been implicated in foodborne outbreaks (Table 1) and should be considered the principal targets for virological analysis. Drinking water, particularly bottled water, also falls into the food category, although to date bottled mineral water has never been clearly identified as a source of viral infections, in spite of reports of the occurrence of NoV sequences (8, 9, 74). However, mineral water may also need to be analyzed for viruses since studies have described cases of contamination with pathogenic bacteria (79, 85).

PROSPECTIVE VERSUS RETROSPECTIVE ANALYSIS OF FOOD

Prospective virological analysis of food is envisaged to ensure the safety of the foodstuff before public consumption. It is such a highly complex and costly process that obviously it cannot be universally applicable to the huge production of the targeted matrices. The first challenge is to choose representative samples; this is even more important when the screening procedures are based on molecular analysis of minute amounts of sample, in which the virus contamination, if present, is very low. Obviously any error at this step is critically carried over and is impossible to correct even when highly sophisticated detection procedures are subsequently applied. Although International Organization for Standardization (ISO) standards for sampling are currently under development, this is still an unresolved issue for molecular viral testing.

A sensible prospective food safety approach is to identify and prevent hazards that could cause food-borne illnesses rather than relying on spot checks of the manufacturing processes and random sampling of finished products to ensure safety. This is the basis of the Hazard Analysis and Critical Control Point (HACCP) system that has been adopted by developed countries (http://ec.europa.eu/food/food/biosafety/hygienelegislation/guide_en.htm; http://www.cfsan.fda.gov/~lrd/haccp.html; http://www.haccp.com.au/). HACCP implementation requires good hygiene practices and good manufacturing practices. Application of the HACCP system to viral safety should be based on the following steps:

- Identify relevant viruses, e.g., NoV and HAV, as hazards and the matrices more susceptible to be contaminated by these viruses.
- Identify critical control points, such as quality of harvesting areas for shellfish or quality of the water employed to irrigate fresh products.
- Establish preventive measures with critical limits for each control point, such as avoiding sewage discharges in shellfish harvesting areas and validating disinfection practices applied to wastewater employed for irrigation.
- Establish procedures based on virus monitoring to verify that the system is working properly.
- Establish corrective actions to take when monitoring shows that a critical limit has not been met. Such actions might include virus analysis of the food product before its commercial distribution.
- Establish effective record keeping.

There is a need to ensure that controls are carried out efficiently and uniformly. It is also vital to instruct third-country staff regarding legal import requirements of receiving countries.

Retrospective analyses of food are applied to isolate the etiological agent diagnosed in the clinical samples in targeted food matrices. Epidemiological case-control studies are necessary to correlate transmission of the disease with exposure to specific food matrices. Two different scenarios may be envisaged. On the one hand, there are situations in which large batches of the same foodstuff linked to the outbreak are still readily available for distribution, where rapid measures to prevent spreading of the outbreak (such as immobilization of food stocks and prohibition of harvesting) may be adopted (7, 75). On the other hand, situations occur in which small amounts of food, mostly associated with improper food handling, are related to the outbreak; these foods are usually not well preserved, if they have been preserved at all, and they fail to be representative of the outbreak cause. The first scenario is worth expending significant efforts to identify and characterize the virus; the second would in most instances be a waste of time and money from the public health point of view. However, it could provide important information from the legal point of view regarding involvement of the foodstuff in transmitting the disease. Evidence for the involvement of food handlers is provided by the isolation of the etiological agent in their stools.

TOWARD STANDARDIZATION OF VIRUS DETECTION PROCEDURES IN FOOD—UTOPIA OR REALITY?

Methods for virus detection cited in the literature are diverse, complex, poorly standardized, and essentially restricted to specialist laboratories. It seems obvious that harmonization of the molecular techniques, as well as addressing quality assurance and quality control (QA/QC) issues, is required before the procedures can be adopted by routine monitoring laboratories. QA/QC measures include the use of positive and negative controls to trace any false-negative or false-positive results, respectively. Most false-negative results are a consequence of inefficient virus and/or nucleic acid extraction and of inhibition of the RT reaction. Most false-positive results are due to cross-contamination.

The first dilemma is to choose between an actual internal control and an added external control for the extraction procedure. For the diagnosis of an RNA virus, the use of an internal control based on detection of the expression of a housekeeping gene, ideally containing introns, through the amplification of its mRNA in the target tissues is a clear first choice. However, this is unrealistic for use in food virology, which involves an increasingly heterogeneous selection of food matrices. For instance, considering only shellfish, a pair of primers to amplify an mRNA for a specific hepatopancreas transcribed gene would be required for each species. It is impossible to apply this criterion to the range of foodstuffs that need to be assayed for viruses; this leads to the use of an external control that is applicable to all matrices under assay. The best candidate for this latter purpose is the use of an encapsidated RNA, such as RNA bacteriophages, RNA animal viruses, or armored RNAs, which are pseudoparticles made of target RNAs packaged into MS2 coliphage coat protein (Table 2) (19, 37, 60, 66). From the purely structural point of view, $T=3$ symmetry particles of around 30 nm in diameter appear to be the most appropriate. Members of the *Leviviridae* family, such as MS2 and f2 coliphages, could fulfill this latter requirement, but they should not be used as standards since they may be present in the target sample (17, 35, 68). In the case of armored RNAs, while it is true that the RNA inside the pseudoparticle is protected from RNases, the MS2 particle coat, although similar to those of NoV and HAV in size, is composed of the CP protein, which has no structural similarity to those of eukaryote isosahedral RNA viruses (27). Since the rationale is to use a nonpathogenic virus of similar structural characteristics to those of the target viruses, nonpathogenic members of the families *Picornaviridae* and *Caliciviridae* could be used to validate the behavior of HAV and NoV, respectively, during the nucleic acid extraction procedures. Among members of the *Picornaviridae*, encephalomyocarditis virus (EMCV) has been proposed as a model for HAV in validation studies of HAV removal in blood product manufacturing by several agencies, such as the European Agency for the Evaluation of Medicinal Products (http://www.emea.eu.int/pdfs/human/bwp/026995en.pdf) and the U.S. Food and Drug Administration (http://www.fda.gov/cber/sba/igivbax042705S .pdf). While the use of EMCV is hampered by its potential pathogenicity in several animals, including primates (18) and even humans (48), the MC_0 strain of Mengo virus, which is serologically indistinguishable from EMCV, is nonpathogenic for humans and other animal species since it lacks the poly(C) tract from the 5′NCR (64); it is used as a vaccine for a wide variety of hosts, including baboons, macaques, and domestic pigs (6). In consequence, Mengo virus MC_0 is being employed as a virus/nucleic acid extrac-

Table 2 QC measures for real-time RT-PCR-based analysis of food

Assay step	Measures
Virus and nucleic acid extractions	Addition of encapsidated RNAs as:
	Bacteriophages
	Animal viruses
	Armored RNA
	Known negative sample
	Reagent negative control
RT-PCR	Addition of RNA molecules
	Addition of dsDNA molecules
	Reagent negative controls

tion control (19). Among *Caliciviridae*, feline calicivirus (FCV) and particularly murine NoV-1 (MNV-1) appear to be good surrogates for human NoV (14) and thus appear to have a use as extraction controls for this virus. However, FCV is pathogenic for cats (63), and hence MNV-1 could be a better option as a NoV standard reagent since it replicates in mouse dendritic cells and macrophages, including the RAW 264.7 cell line (86, 87). The problem with using MNV-1 as a standard reagent is its low genomic robustness, since, as stated above, NoV RNAs apparently lack functional, highly structured elements that make ideal targets for quantitative assays.

The requirement for the addition of some kind of internal controls for follow-up of the molecular reactions, particularly the RT, has been suggested by many authors (1, 4, 5, 19, 39, 42, 43, 65). The control of an RT-PCR assay should rely on a single-stranded RNA (ssRNA) molecule that corresponds to the viral target and thus is amplifiable with the same pair of primers under exactly the same conditions and with the same efficiency. Ideally, the size of this RNA should be as similar as possible to that of the target, and the RNA should be easily differentiated from the target.

Finally, another issue is the choice of the type of molecule to be used in generating the standard curve of the target quantification. Since virus nucleic acid extraction and RT-PCR efficiencies are already measured by the use of the aforementioned specific control reagents, and bearing in mind that real-time PCR is a method for DNA quantification, a double-stranded (dsDNA) molecule corresponding to the amplimer (e.g., the actual amplimer or a larger dsDNA molecule containing the amplimer) is the optimal choice. A best-case scenario is set for the standard curve, since otherwise the corrective coefficients of the critical assay steps would be overestimated.

The use of this set of controls allows the use of QA/QC measures that enable a correct interpretation of the data, uncovering false-negative and false-positive results. A quantitative assay additionally involves the generation of three different standard curves: one for the quantification of the chosen encapsidated RNA, another for the quantification of the ssRNA molecule control, and a third for the quantification of the virus target, made with the dsDNA molecule, thus providing the possibility of accurate quantification after applying the corrective factors derived by the efficiencies of the extraction and RT steps (Fig. 3). An accurate quantification requires answering additional questions such as determining acceptable efficiency thresh-

---

Figure 3 Proposed standardized procedure for an accurate estimation of the number of viral genome copies in food. (Reprinted from reference 19 with permission.)

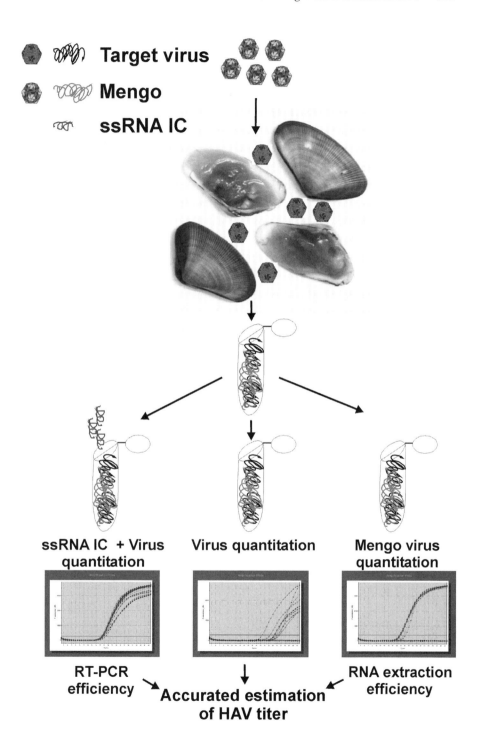

olds, particularly regarding the virus-nucleic acid extraction procedures, and establishing adequate uncertainty levels of detection for the assays. It is worth mentioning that these approaches are followed by the CEN/TC 275/ WG6/TAG4 Committee (European Committee for Standardization of Horizontal Methods for the Molecular Detection of Viruses in Food) to develop ISO proposals for sensitive and quantitative RT-PCR-based methods to detect HAV and NoV in selected commodities; this could enable the formulation of regulatory standards for detection of viruses in food.

Another unsolved issue in food virology is the significance of genome copies in terms of infectivity. The lack of cell culture systems applicable to NoV and wild-type HAV propagation has long prevented the development of infectivity assays for these agents. The recently described cell culture systems for HAV and NoV (50, 77) may make it possible to address this problem. Until the adequacy of these new infectivity assays, either alone or combined with molecular approaches, is validated, other measures to increase the significance of genome copy numbers, such as pretreating the sample with RNase, could be used (62). It is generally accepted that all RNA viruses have a low specific infectivity or infectious/physical particle ratio (10). This low ratio has been suggested to be due to lethal genetic defects and/or to unsuccessful infectious cycles (57, 71). However, these noninfectious unaltered RNA-containing capsids may still be detected by molecular techniques even after RNase pretreatment.

The advent of standardized molecular techniques allowing the detection and quantification of viruses in foodstuffs is a major breakthrough in food safety. However, despite this progress, there is still a long way to go before these techniques could be applied on a routine basis to significant food samples, which would be particularly relevant in the present situation of global food trade. Nevertheless, once these standardized virological assays become available, the formulation of guidelines to ensure the virological quality of selected commodities in specific scenarios will certainly contribute to reducing the incidence of food-borne virus infections.

REFERENCES

1. **Abid, I., S. Guix, M. Aouni, R. Pinto, and A. Bosch.** 2007. Detection and characterization of human group C rotavirus in the pediatric population of Barcelona, Spain. *J. Clin. Virol.* **38:**78–82.

2. **Abuzwaida, A. R. N., M. Sidoni, C. F. T. Yoshida, and H. G. Schatzmayr.** 1987. Seroepidemiology of hepatitis-A and hepatitis-B in 2 urban communities of Rio-de-Janeiro, Brazil. *Rev. Inst. Med. Trop. Sao Paulo* **29:**219–223.

3. **Asanaka, M., R. L. Atmar, V. Ruvolo, S. E. Crawford, F. H. Neill, and M. K. Estes.** 2005. Replication and packaging of Norwalk virus RNA in cultured mammalian cells. *Proc. Natl. Acad. Sci. USA.* **102:**10327–10332.

4. **Atmar, R. L.** 2006. Molecular methods of virus detection in foods, p. 121–149. *In* S. M. Goyal (ed.), *Viruses in Foods.* Springer, New York, NY.

5. **Atmar, R. L., and M. K. Estes.** 2001. Diagnosis of noncultivatable gastroenteritis viruses, the human caliciviruses. *Clin. Microbiol. Rev.* **14:**15–37.

6. **Backues, K. A., M. Hill, A. C. Palmenberg, C. Miller, K. F. Soike, and R. Aguilar.** 1999. Genetically engineered mengo virus vaccination of multiple captive wildlife species. *J. Wildl. Dis.* **35:**384–387.

7. **Berg, D. E., M. A. Kohn, T. A. Farley, and L. M. McFarland.** 2000. Multi-state outbreaks of acute gastroenteritis traced to fecal-contaminated oysters harvested in Louisiana. *J. Infect. Dis.* **181**(Suppl. 2)**:**381–386.

8. **Beuret, C., D. Kohler, A. Baumgartner, and T. M. Luthi.** 2002. Norwalk-like virus sequences in mineral waters: one-year monitoring of three brands. *Appl. Environ. Microbiol.* **68:**1925–1931.

9. **Beuret, C., D. Kohler, and T. Luthi.** 2000. Norwalk-like virus sequences detected by reverse transcription-polymerase chain reaction in mineral waters imported into or bottled in Switzerland. *J. Food Prot.* **63:**1576–1582.

10. **Biebricher, C. K., and M. Eigen.** 2005. The error threshold. *Virus Res.* **107:**117–127.

11. **Bostock, A. D., P. Mepham, S. Phillips, S. Skidmore, and M. H. Hambling.** 1979. Hepatitis A infection associated with the consumption of mussels. *J. Infect.* **1:**171–177.

12. **Brown, E. A., A. J. Zajac, and S. M. Lemon.** 1994. In vitro characterization of an internal ribosomal entry site (IRES) present within the 5′ nontranslated region of hepatitis A virus RNA: comparison with the IRES of encephalomyocarditis virus. *J. Virol.* **68:**1066–1074.

13. **Calder, L., G. Simmons, C. Thornley, P. Taylor, K. Pritchard, G. Greening, and J. Bishop.** 2003. An outbreak of hepatitis A associated with consumption of raw blueberries. *Epidemiol. Infect.* **131:**745–751.

14. **Cannon, J. L., E. Papafragkou, G. W. Park, J. Osborne, L. A. Jaykus, and J. Vinje.** 2006. Surrogates for the study of norovirus stability and inactivation in the environment: a comparison of murine norovirus and feline calicivirus. *J. Food Prot.* **69:**2761–2765.

15. **Centers for Disease Control and Prevention.** 1997. Hepatitis A associated with consumption of frozen strawberries—Michigan, March 1997. *Morb. Mortal. Wkly. Rep.* **46:**288–295.

16. **Chen, R., J. D. Neill, J. S. Noel, A. M. Hutson, R. I. Glass, M. K. Estes, and B. V. V. Prasad.** 2004. Inter- and intragenus structural variations in caliciviruses and their functional implications. *J. Virol.* **78:**6469–6479.

17. **Chung, H., L. A. Jaykus, G. Lovelace, and M. D. Sobsey.** 1998. Bacteriophages and bacteria as indicators of enteric viruses in oysters and their harvest waters. *Water Sci. Technol.* **38:**37–44.

18. **Citino, S. B., B. L. Homer, J. H. Gaskin, and D. J. Wickham.** 1988. Fatal encephalomyocarditis virus infection in a Sumatran orangutan (*Pongo pygmaeus abelii*). *J. Zoo Wildl. Med.* **19:**214–218.

19. **Costafreda, M. I., A. Bosch, and R. M. Pinto.** 2006. Development, evaluation, and standardization of a real-time TaqMan reverse transcription-PCR assay for quantification of hepatitis A virus in clinical and shellfish samples. *Appl. Environ. Microbiol.* **72:**3846-3855.

20. Costa-Mattioli, M., J. Cristina, H. Romero, R. Perez-Bercof, D. Casane, R. Colina, L. Garcia, I. Vega, G. Glikman, V. Romanowsky, A. Castello, E. Nicand, M. Gassin, S. Billaudel, and V. Ferre. 2002. Molecular evolution of hepatitis A virus: a new classification based on the complete VP1 protein. *J. Virol.* **76**:9516–9525.

21. Costa-Mattioli, M., S. Monpoeho, E. Nicand, M. H. Aleman, S. Billaudel, and V. Ferre. 2002. Quantification and duration of viraemia during hepatitis A infection as determined by real-time RT-PCR. *J. Viral Hepat.* **9**:101–106.

22. Daemer, R. J., S. M. Feinstone, I. D. Gust, and R. H. Purcell. 1981. Propagation of hepatitis A virus in African green monkey kidney cell culture: primary isolation and serial passage. *Infect. Immun.* **32**:388–393.

23. Dentinger, C. M., W. A. Bower, O. V. Nainan, S. M. Cotter, G. Myers, L. M. Dubusky, S. Fowler, E. D. Salehi, and B. P. Bell. 2001. An outbreak of hepatitis A associated with green onions. *J. Infect. Dis.* **183**:1273–1276.

24. Domingo, E., and J. J. Holland. 1997. RNA virus mutations and fitness for survival. *Annu. Rev. Microbiol.* **51**:151–178.

25. Duizer, E., K. J. Schwab, F. H. Neill, R. L. Atmar, M. P. Koopmans, and M. K. Estes. 2004. Laboratory efforts to cultivate noroviruses. *J. Gen. Virol.* **85**:79–87.

26. Ehrenfeld, E., and N. L. Teterina. 2002. Initiation of translation of picornavirus RNAs: structure and function of the internal ribosome entry site, p. 159–169. *In* B. L. W. E. Semler (ed.), *Molecular Biology of Picornaviruses*. ASM Press, Washington, DC.

27. Fauquet, C. M., M. A. Mayo, J. Maniloff, U. Desselberger, and L. A. Ball. 2005. *Virus Taxonomy. Eighth Report of the International Committee on Taxonomy of Viruses.* Elsevier Academic Press, New York, NY.

28. Fix, A. D., M. Abdel-Hamid, R. H. Purcell, M. H. Shehata, F. Abdel-Aziz, N. Mikhail, H. el Sebai, M. Nafeh, M. Habib, R. R. Arthur, S. U. Emerson, and G. T. Strickland. 2000. Prevalence of antibodies to hepatitis E in two rural Egyptian communities. *Am. J. Trop. Med. Hyg.* **62**:519–523.

29. Flehmig, B. 1980. Hepatitis A virus in cell culture. 1. Propagation of different hepatitis A isolates in a foetal rhesus monkey kidney cell line (FRhK-4). *Med. Microbiol. Immunol.* **168**:239–248.

30. Fukuda, S., S. Takao, M. Kuwayama, Y. Shimazu, and K. Miyazaki. 2006. Rapid detection of norovirus from fecal specimens by real-time reverse transcription-loop-mediated isothermal amplification assay. *J. Clin. Microbiol.* **44**:1376–1381.

31. Gaspar, A. M., C. L. Vitral, R. S. Marchevsky, C. F. T. Yoshida, and H. G. Schatzmayr. 1992. A Brazilian hepatitis-A virus isolated and adapted in primate and primate cell-line as a chance for the development of a vaccine. *Mem. Inst. Oswaldo Cruz* **87**:449–450.

32. Gaulin, C. D., D. Ramsay, P. Cardinal, and M. A. D'Halevyn. 1999. Epidemic of gastroenteritis of viral origin associated with eating imported raspberries. *Can. J. Public Health* **90**:37–40. (In French.)

33. Green, K. Y., R. M. Chanock, and A. Z. Kapikian. 2001. Human caliciviruses, p. 841–874. *In* B. N. Fields, D. M. Knipe, P. M. Howley, and D. E. Griffin (ed.), *Fields Virology*, 4th ed. Lippincott Williams & Wilkins, Philadelphia, PA.

34. Halliday, M. L., L.-Y. Kang, T.-Z. Zhou, M.-D. Hu, Q.-C. Pan, T.-Y. Fu, Y. S. Huang, and S.-L. Hu. 1991. An epidemic of hepatitis A attributable to the ingestion of raw clams in Shanghai, China. *J. Infect. Dis.* **164**:852–859.

35. **Havelaar, A. H., K. Furuse, and W. M. Hogeboom.** 1986. Bacteriophages and indicator bacteria in human and animal faeces. *J. Appl. Bacteriol.* **60:**255–262.

36. **Hewitt, J., and G. E. Greening.** 2004. Survival and persistence of norovirus, hepatitis A virus, and feline calicivirus in marinated mussels. *J. Food Prot.* **67:**1743–1750.

37. **Hietala, S. K., and B. M. Crossley.** 2006. Armored RNA as virus surrogate in a real-time reverse transcriptase PCR assay proficiency panel. *J. Clin. Microbiol.* **44:**67–70.

38. **Hohne, M., and E. Schreier.** 2004. Detection and characterization of norovirus outbreaks in Germany: application of a one-tube RT-PCR using a fluorogenic real-time detection system. *J. Med. Virol.* **72:**312–319.

39. **Hoorfar, J., B. Malorny, A. Abdulmawjood, N. Cook, M. Wagner, and P. Fach.** 2004. Practical considerations in design of internal amplification controls for diagnostic PCR assays. *J. Clin. Microbiol.* **42:**1863–1868.

40. **Jean, J., B. Blais, A. Darveau, and I. Fliss.** 2001. Detection of hepatitis A virus by the nucleic acid sequence-based amplification technique and comparison with reverse transcription-PCR. *Appl. Environ. Microbiol.* **67:**5593–5600.

41. **Jean, J., D. H. D'Souza, and L. A. Jaykus.** 2004. Multiplex nucleic acid sequence-based amplification for simultaneous detection of several enteric viruses in model ready-to-eat foods. *Appl. Environ. Microbiol.* **70:**6603–6610.

42. **Jothikumar, N., T. L. Cromeans, M. D. Sobsey, and B. H. Robertson.** 2005. Development and evaluation of a broadly reactive TaqMan assay for rapid detection of hepatitis A virus. *Appl. Environ. Microbiol.* **71:**3359–3363.

43. **Jothikumar, N., J. A. Lowther, K. Henshilwood, D. N. Lees, V. R. Hill, and J. Vinje.** 2005. Rapid and sensitive detection of noroviruses by using TaqMan-based one-step reverse transcription-PCR assays and application to naturally contaminated shellfish samples. *Appl. Environ. Microbiol.* **71:**1870–1875.

44. **Kageyama, T., S. Kojima, M. Shinohara, K. Uchida, S. Fukushi, F. B. Hoshino, N. Takeda, and K. Katayama.** 2003. Broadly reactive and highly sensitive assay for Norwalk-like viruses based on real-time quantitative reverse transcription-PCR. *J. Clin. Microbiol.* **41:**1548–1557.

45. **Kageyama, T., M. Shinohara, K. Uchida, S. Fukushi, F. B. Hoshino, S. Kojima, R. Takai, T. Oka, N. Takeda, and K. Katayama.** 2004. Coexistence of multiple genotypes, including newly identified genotypes, in outbreaks of gastroenteritis due to norovirus in Japan. *J. Clin. Microbiol.* **42:**2988–2995.

46. **Katayama, H., A. Shimasaki, and S. Ohgaki.** 2002. Development of a virus concentration method and its application to detection of enterovirus and Norwalk virus from coastal seawater. *Appl. Environ. Microbiol.* **68:**1033–1039.

47. **Katayama, K., H. Shirato-Horikoshi, S. Kojima, T. Kageyama, T. Oka, F. Hoshino, S. Fukushi, M. Shinohara, K. Uchida, Y. Suzuki, T. Gojobori, and N. Takeda.** 2002. Phylogenetic analysis of the complete genome of 18 Norwalk-like viruses. *Virology* **299:**225–239.

48. **Kirkland, P. D., R. A. Hawkes, H. M. Naim, and C. R. Boughton.** 1989. Human infection with encephalomyocarditis virus in New South Wales. *Med. J. Aust.* **151:**176.

49. **Kohn, M. A., T. A. Farley, T. Ando, M. Curtis, S. A. Wilson, Q. Jin, S. S. Monroe, R. C. Baron, L. M. McFarland, and R. I. Glass.** 1995. An outbreak of Norwalk virus gastroenteritis associated with eating raw oysters. Implications for maintaining safe oyster beds. *JAMA* **273:**466–471.

50. Konduru, K., and G. Kaplan. 2006. Stable growth of wild-type hepatitis a virus in cell culture. *J. Virol.* **80:**1352–1360.

51. Loisy, F., R. L. Atmar, P. Guillon, P. Le Cann, M. Pommepuy, and F. S. Le Guyader. 2005. Real-time RT-PCR for norovirus screening in shellfish. *J. Virol. Methods* **123:**1–7.

52. Lopman, B., H. Vennema, E. Kohli, P. Pothier, A. Sanchez, A. Negredo, J. Buesa, E. Schreier, M. Reacher, D. Brown, J. Gray, M. Iturriza, C. Gallimore, B. Bottiger, K. O. Hedlund, M. Torven, C. H. von Bonsdorff, L. Maunula, M. Poljsak-Prijatelj, J. Zimsek, G. Reuter, G. Szucs, B. Melegh, L. Svennson, Y. van Duijnhoven, and M. Koopmans. 2004. Increase in viral gastroenteritis outbreaks in Europe and epidemic spread of new norovirus variant. *Lancet* **363:**682–688.

53. Lopman, B. A., M. H. Reacher, Y. van Duijnhoven, F. X. Hanon, D. Brown, and M. Koopmans. 2003. Viral gastroenteritis outbreaks in Europe, 1995–2000. *Emerg. Infect. Dis.* **9:**90–96.

54. Luttermann, C., and G. Meyers. 2007. A bipartite sequence motif induces translation reinitiation in feline calicivirus RNA. *J. Biol. Chem.* **282:**7056–7065.

55. Martin, A., and S. M. Lemon. 2006. Hepatitis A virus: from discovery to vaccines. *Hepatology* **43:**S164–S172.

56. Mead, P. S., L. Slutsker, V. Dietz, L. F. McCaig, J. S. Bresee, C. Shapiro, P. M. Griffin, and R. V. Tauxe. 1999. Food-related illness and death in the United States. *Emerg. Infect. Dis.* **5:**607–625.

57. Mueller, S., D. Papamichail, J. R. Coleman, S. Skiena, and E. Wimmer. 2006. Reduction of the rate of poliovirus protein synthesis through large-scale codon deoptimization causes attenuation of viral virulence by lowering specific infectivity. *J. Virol.* **80:**9687–9696.

58. Nainan, O. V., G. L. Armstrong, X. H. Han, I. Williams, B. P. Bell, and H. S. Margolis. 2005. Hepatitis A molecular epidemiology in the United States, 1996–1997: sources of infection and implications of vaccination policy. *J. Infect. Dis.* **191:**957–963.

59. Nainan, O. V., G. L. Xia, G. Vaughan, and H. S. Margolis. 2006. Diagnosis of hepatitis a virus infection: a molecular approach. *Clin. Microbiol. Rev.* **19:**63–79.

60. Nishida, T., H. Kimura, M. Saitoh, M. Shinohara, M. Kato, S. Fukuda, T. Munemura, T. Mikami, A. Kawamoto, M. Akiyama, Y. Kato, K. Nishi, K. Kozawa, and O. Nishio. 2003. Detection, quantitation, and phylogenetic analysis of noroviruses in Japanese oysters. *Appl. Environ. Microbiol.* **69:**5782–5786.

61. Notomi, T., H. Okayama, H. Masubuchi, T. Yonekawa, K. Watanabe, N. Amino, and T. Hase. 2000. Loop-mediated isothermal amplification of DNA. *Nucleic Acids Res.* **28:**e63.

62. Nuanualsuwan, S., and D. O. Cliver. 2002. Pretreatment to avoid positive RT-PCR results with inactivated viruses. *J. Virol. Methods* **104:**217–225.

63. Ormerod, E., I. A. P. Mccandlish, and O. Jarrett. 1979. Diseases produced by feline caliciviruses when administered to cats by aerosol or intranasal instillation. *Vet. Rec.* **104:**65–69.

64. Osorio, J. E., G. B. Hubbard, K. F. Soike, M. Girard, S. van der Werf, J. C. Moulin, and A. C. Palmenberg. 1996. Protection of non-murine mammals against encephalomyocarditis virus using a genetically engineered Mengo virus. *Vaccine* **14:**155–161.

65. Parshionikar, S. U., J. L. Cashdollar, and G. S. Fout. 2004. Development of homologous viral internal controls for use in RT-PCR assays of waterborne enteric viruses. *J. Virol. Methods* **121:**39–48.

66. **Pasloske, B. L., C. R. Walkerpeach, R. D. Obermoeller, M. Winkler, and D. B. DuBois.** 1998. Armored RNA technology for production of ribonuclease-resistant viral RNA controls and standards. *J. Clin. Microbiol.* **36:**3590–3594.

67. **Paul, A. V.** 2002. Possible unifying mechanism of picornavirus genome replication, p. 227–246. *In* B. L. Semler and E. Wimmer (ed.), *Molecular Biology of Picornaviruses.* ASM Press, Washington, DC.

68. **Pillai, S. D.** 2006. Bacteriophages as fecal indicator organisms, p. 205–222. *In* S. M. Goyal (ed.), *Viruses in Foods.* Springer, New York, NY.

69. **Ponka, A., L. Maunula, C. H. von Bonsdorff, and O. Lyytikainen.** 1999. An outbreak of calicivirus associated with consumption of frozen raspberries. *Epidemiol. Infect.* **123:**469–474.

70. **Poyry, T., L. Kinnunen, and T. Hovi.** 1992. Genetic variation in vivo and proposed functional domains of the 5′ noncoding region of poliovirus RNA. *J. Virol.* **66:**5313–5319.

71. **Racaniello, V. R.** 2001. Picornaviridae: the viruses and their replication, p. 685–722. *In* B. N. Fields, D. M. Knipe, P. M. Howley, and D. E. Griffin (ed.), *Fields Virology,* 4th ed. Lippincott Williams & Wilkins, Philadelphia, PA.

72. **Robertson, B. H., R. W. Jansen, B. Khanna, A. Totsuka, O. V. Nainan, G. Siegl, A. Widell, H. S. Margolis, S. Isomura, K. Ito, T. Ishizu, Y. Moritsugu, and S. M. Lemon.** 1992. Genetic relatedness of hepatitis A virus strains recovered from different geographical regions. *J. Gen. Virol.* **73:**1365–1377.

73. **Rosenblum, L. S., I. R. Mirkin, D. T. Allen, S. Safford, and S. C. Hadler.** 1990. A multifocal outbreak of hepatitis A traced to commercially distributed lettuce. *Am. J. Public Health* **80:**1075–1079.

74. **Sanchez, G., H. Joosten, R. Meyer, and C. Beuret.** 2005. Presence of norovirus sequences in bottled waters is questionable. *Appl. Environ. Microbiol.* **71:**2203–2205.

75. **Sanchez, G., R. M. Pinto, H. Vanaclocha, and A. Bosch.** 2002. Molecular characterization of hepatitis a virus isolates from a transcontinental shellfish-borne outbreak. *J. Clin. Microbiol.* **40:**4148–4155.

76. **Shieh, Y. C., Y. E. Khudyakov, G. Xia, L. M. Ganova-Raeva, F. M. Khambaty, J. W. Woods, J. E. Veazey, M. L. Motes, M. B. Glatzer, S. R. Bialek, and A. E. Fiore.** 2007. Molecular confirmation of oysters as the vector for hepatitis A in a 2005 multistate outbreak. *J. Food Prot.* **70:**145–150.

77. **Straub, T. M., K. Höner zu Bentrup, P. O. Orosz-Coghlan, A. Dohnalkova, B. K. Mayer, R. A. Bartholomew, C. O. Valdez, C. J. Bruckner, C. P. Gerba, M. Abbaszadegan, and C. A. Nickerson.** 2007. In vitro cell culture infectivity assay for human noroviruses. *Emerg. Infect. Dis.* **13:**396–403.

78. **Tallo, T., H. Norder, V. Tefanova, K. Ott, V. Ustina, T. Prukk, O. Solomonova, J. Schmidt, K. Zilmer, L. Priimägi, T. Krispin, and L. O. Magnius.** 2003. Sequential changes in hepatitis A virus genotype distribution in Estonia during 1994 to 2001. *J. Med. Virol.* **70:**187–193.

79. **Tsai, G. J., and S. C. Yu.** 1997. Microbiological evaluation of bottled uncarbonated mineral water in Taiwan. *Int. J. Food Microbiol.* **37:**137–143.

80. **Vennema, H., E. de Bruin, and M. Koopmans.** 2002. Rational optimization of generic primers used for Norwalk-like virus detection by reverse transcriptase polymerase chain reaction. *J. Clin. Virol.* **25:**233–235.

81. Vinjé, J., J. Green, D. C. Lewis, C. I. Gallimore, D. W. Brown, and M. P. Koopmans. 2000. Genetic polymorphism across regions of the three open reading frames of "Norwalk-like viruses." *Arch. Virol.* **145**:223–241.

82. Vinjé, J., H. Vennema, L. Maunula, C. H. von Bonsdorff, M. Hoehne, E. Schreier, A. Richards, J. Green, D. Brown, S. S. Beard, S. S. Monroe, E. de Bruin, L. Svensson, and M. P. Koopmans. 2003. International collaborative study to compare reverse transcriptase PCR assays for detection and genotyping of noroviruses. *J. Clin. Microbiol.* **41**:1423–1433.

83. Vinjé, J., R. A. Hamidjaja, and M. D. Sobsey. 2004. Development and application of a capsid VP1 (region D) based reverse transcription PCR assay for genotyping of genogroup I and II noroviruses. *J. Virol. Methods* **116**:109–117.

84. Wheeler, C., T. M. Vogt, G. L. Armstrong, G. Vaughan, A. Weltman, O. V. Nainan, V. Dato, G. L. Xia, K. Waller, J. Amon, T. M. Lee, A. Highbaugh-Battle, C. Hembree, S. Evenson, M. A. Ruta, I. T. Williams, A. E. Fiore, and B. P. Bell. 2005. An outbreak of hepatitis A associated with green onions. *N. Engl. J. Med.* **353**:890–897.

85. Wilkinson, F. H., and K. G. Kerr. 1998. Bottled water as a source of multi-resistant *Stenotrophomonas* and *Pseudomonas* species for neutropenic patients. *Eur. J. Cancer Care* **7**:12–14.

86. Wobus, C. E., S. M. Karst, L. B. Thackray, K. O. Chang, S. V. Sosnovtsev, G. Belliot, A. Krug, J. M. Mackenzie, K. Y. Green, and H. W. Virgin. 2004. Replication of Norovirus in cell culture reveals a tropism for dendritic cells and macrophages. *PLoS Biol.* **2**:2076–2084.

87. Wobus, C. E., L. B. Thackray, and H. W. Virgin. 2006. Murine norovirus: a model system to study norovirus biology and pathogenesis. *J. Virol.* **80**:5104–5112.

Food-Borne Viruses: Progress and Challenges
Edited by Marion P. G. Koopmans, Dean O. Cliver, and Albert Bosch
© 2008 ASM Press, Washington, DC

Binding and Inactivation of Viruses on and in Food, with a Focus on the Role of the Matrix

8

Françoise S. Le Guyader and Robert L. Atmar

Foods play an important role in the transmission of enteric viruses. For example, as many as 40% of norovirus (NoV) infections are estimated to result from the consumption of contaminated foodstuffs (52, 67). Although fewer than 5% of hepatitis A virus (HAV) infections are classified as food borne, a greater percentage may be transmitted by this route since the source of infection is not identified in up to one-half of cases (3, 19, 53). Other enteric viruses, including rotaviruses (see chapter 2), astroviruses (see chapter 5), and hepatitis E virus (see chapter 3), are not important causes of food-borne disease but are occasionally transmitted by foods (12, 13, 61, 74, 98), and nonenteric viruses are occasionally also transmitted in foods (e.g., tick-borne encephalitis virus and Nipah virus [see chapter 5]) (44, 63).

Transmission of viruses by foods is dependent on several factors: initial contact of the virus with the food, binding or attachment of the virus to the food, survival and persistence of the virus until the food is consumed, and ingestion of the food by a susceptible host. The initial contact of the virus with the food may occur at any time during food production, including before harvest, during processing, and at the time of preparation (Fig. 1). Examples include fecal contamination of shellfish-growing waters due to poorly functioning sewage treatment plants, use of night soil to fertilize crops, use of fecally contaminated water to wash fruits after harvest, and poor hand hygiene by an infected food handler (8, 12, 52, 86, 97).

FRANÇOISE S. LE GUYADER, Laboratoire de Microbiologie, IFREMER, BP21105, Nantes 44311, France. ROBERT L. ATMAR, Departments of Medicine and Molecular Virology & Microbiology, Baylor College of Medicine, 1 Baylor Plaza, Houston, TX 77030.

Figure 1 Environmental sources of contamination for food (vegetables and shellfish). Food may be contaminated directly by sewage, rivers, or fertilizers. Climate events such as rain, sunshine, and temperature may affect virus behavior before harvesting.

Almost any kind of food can be involved in virus transmission, but a limited number of foods are most commonly associated with outbreaks. These include uncooked shellfish, fresh fruits and vegetables, and ready-to-eat foods (such as sandwiches and salads). Ready-to-eat foods are often contaminated by an infected food handler at the time of food preparation, while contamination of shellfish usually occurs before harvesting, and contamination of fruits and vegetables can occur before harvesting or during processing. This chapter reviews factors that are known to affect the binding of enteric viruses to different food matrices and that affect the persistence and survival of the virus once it is food associated, with an emphasis on foods that are contaminated by food handlers before the preparation stage.

VIRUS BINDING TO FOODS

Physicochemical Factors

The adsorption of viruses to solid surfaces has been evaluated for a number of different surface types (sorbents), including soil and aquifer sediments, food preparation surfaces, and fomites (2, 32, 42, 89). The size of viruses is similar to that of colloids; therefore, theories originally developed to describe the binding of colloids to surfaces have been used to model virus binding to different sorbents. The Derjaguin-Landau-Verwey-Overbeek (DLVO) theory of colloidal stability states that adherence of a colloid to a surface results from the interaction of two opposing forces: attractive van der Waals forces and repulsive electrostatic forces. Factors that affect the electrostatic forces include the pH and ionic strength of the solution, the presence of compounds competing for sorption sites, properties of the virus (e.g., its isoelectric point [pI]), and properties of the sorbent (42) (Fig. 2). For example, at pH values below the pI the net charge of the virus particle is positive, while at pH values above the pI, the net charge is negative. Thus, changes in pH affect the electrostatic interactions with the sorbent, depending on its net charge. This characteristic has been used to elute viruses from sorbents by increasing the pH of the solution bathing the sorbent to >9.0. Increasing the ionic strength of a solution favors adsorption of a virus to a sorbent by decreasing the effective radius of repulsive electrostatic effects. The converse is also true—lowering the ionic strength of the solution is used to elute virus from a surface (32). Experimental results from studies of poliovirus binding to different metal oxides have agreed with predictions by the DLVO model (42).

The DLVO theory does not consider the contribution of hydrophobic interactions between virus and sorbent, and at least for some viruses (e.g., bacteriophages) used to model enteric virus interactions, hydrophobic

Figure 2 Factors influencing the probability that viruses will bind to food.

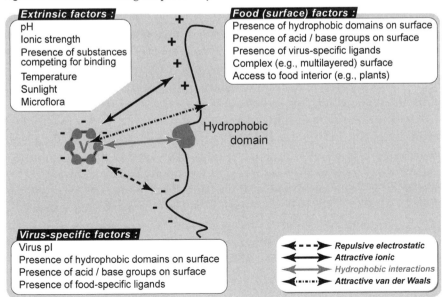

interactions contribute significantly to the adsorption of the viruses to selected sorbents (15). Relatively few studies have evaluated the adsorption of viruses to different foods, but Vega et al. (93) studied the binding of echovirus 11 (used to model picornaviruses) and feline calicivirus (used to model human caliciviruses) to lettuce and found that these viruses did not have adsorption patterns that fit those predicted by the DLVO theory. In contrast, many studies have examined factors affecting the elution of viruses from different foods, and the same variables (e.g., pH, ionic strength, cation concentration) that are important in virus binding to nonfood surfaces also significantly influence virus elution (10, 27, 56, 98) (Fig. 2). The performance of elution buffers is not exactly the same on two types of berries (raspberries and strawberries), suggesting that these two sorbents have different binding characteristics (10). However, no standard procedure or systematic approach to evaluating the adsorption of enteric viruses to different substrates (including foods) has been developed, making it difficult to draw conclusions about the mechanisms involved in virus sorption (42).

Virus Contamination of Plants

Enteric viruses can cause surface contamination of fruits and vegetables by adsorption following exposure to fecally contaminated soil or groundwater (43). A study from Costa Rica used a simple wash to demonstrate the presence of rotavirus and HAV on the surface of lettuce bought at the market (36). Such surface contamination conceivably could be amenable to virus removal by elution. Another possible route of contamination is uptake of virus through the root system and subsequent transport of the virus into edible portions of the plant via the vascular tissue (phloem and xylem). This route of microbial penetration into plants is known to occur with pathogenic plant viruses and has also been noted to occur with pathogenic *Salmonella* strains (35, 43). A limited number of studies have addressed this question for enteric viruses that are pathogenic for humans.

An early study demonstrated that a poliovirus type 1 strain was absorbed by tomato plant roots but was not translocated to aerial parts of the plant, while a mouse encephalomyelitis virus (another picornavirus) was also absorbed in significant concentrations and was occasionally translocated acropetally (71). Polioviruses were also occasionally demonstrable in the upper parts of plants that had been irrigated with virus-contaminated waters (43, 66). Katzenelson and Mills (43) demonstrated that polioviruses most easily gained access to the upper part of pepper plants following damage to the stem but that this also occurred after damage to the root system. Interestingly, no translocation of virus to the upper parts of the pepper plant (or to those of cucumber or lettuce plants) occurred when the root systems of intact plants were exposed to virus.

Because of the recent association of HAV infections with green onion consumption, Chancellor et al. (14) evaluated the uptake of fluorescent microspheres (1 and 10 μm in diameter) and attenuated HAV (a vaccine strain) into growing green onions. The fluorescent beads could be demonstrated within the onion as early as 1 day after exposure, and their number increased over time, while HAV was detectable within the onion by RT-PCR methods 1 week after exposure. While these studies did not address the viability of the virus, they suggest that virus can be taken up into the onion by a process that would prevent its removal by simple cleaning of the onion's surface and may suggest a mechanism for contamination of green onions involved in several hepatitis A outbreaks (14, 23, 96).

Many edible fruits and vegetables have complex surfaces that prevent the removal of contaminating substances by simple washing. For example, washing does not substantially reduce viral titers of HAV from lettuce, fennel, and carrots after the surfaces of these vegetables have been exposed to a solution containing HAV (18). Similar results were observed when MS2 bacteriophage was used to contaminate the surfaces of a variety of fresh fruits and vegetables, including cucumbers, tomatoes, peppers, lettuce, strawberries, parsley, green onions, carrots, and cabbage (21). The complex surfaces of these foods may also decrease desiccation effects that lead to virus inactivation (81).

Virus Uptake by Shellfish and Specific Binding to Tissues

Shellfish are filter feeders and concentrate enteric viruses from their environment while feeding. The majority of accumulated virus is found in the pancreatic tissue, also called the digestive diverticula (85). Virus accumulation in oysters is affected by factors such as water temperature, mucus production, glycogen content of the connective tissue, and gonadal development (9). Mechanical entrapment and ionic bonding are among the mechanisms that have been suggested to explain observed differences in accumulation of different viruses and among different oyster species (9, 26, 68).

Depuration is a dynamic process whereby shellfish purge themselves of contaminants either in a natural setting or in land-based facilities. Relaying is the practice of transferring shellfish harvested from contaminated areas to clean shellfish-growing waters. Both depuration and relaying are effective approaches to removing bacterial enteric pathogens from contaminated shellfish (79). If shellfish acted as mere filters or ionic traps to concentrate viruses, a simple depuration process should be sufficient to rid oysters of virus, as has been observed for bacteria (79). However, numerous studies have shown that oysters eliminate viruses much less efficiently than they eliminate bacteria (79, 85). For example, oysters eliminated only ~7% of Norwalk virus compared to 95% of *Escherichia coli* over a 48-h depuration period (85). Poor depuration efficiency has also been seen for other viruses, including adenovirus, rotavirus, and HAV (37, 47, 62).

Another potential mechanism for the uptake and concentration of viruses in shellfish has been proposed based on the observation of specific binding of a genogroup I NoV to shellfish tissues (58). The genus *Norovirus* is divided into genogroups (I through V) and further subdivided into genotypes (from 1 to 20 genotypes per genogroup) (100). Distinct NoV strains belonging to both genogroups I and II exhibit various binding patterns with different carbohydrate structures of the histo-blood group antigen (HBGA) family, and lack of expression of these carbohydrates has been correlated with resistance to infection in humans (see chapter 2) (4, 90). In vitro expression of the viral structural proteins, VP1 and VP2, leads to the spontaneous formation of virus-like particles (VLPs), and the VLPs are used to model viral interactions. For example, no difference in tissue distribution or binding is seen when Norwalk virus VLPs are compared to Norwalk virus in bioaccumulation experiments and in experiments of binding to *Crassostrea gigas* tissue (58).

An HBGA A-like carbohydrate, α-linked *N*-acetylgalactosamine, is a ligand present in shellfish tissues that is involved in the specific binding of Norwalk virus VLPs to these tissues. The specificity of the binding as an HBGA A-like carbohydrate was demonstrated by using HBGA-specific carbohydrate antibodies and lectins to block VLP binding to tissues. HBGAs present in saliva also block VLP binding, and periodate treatment, which removes carbohydrate residues, decreases VLP binding to tissues (58). These observations are similar to those made during VLP-binding studies with human intestinal tissues (41, 90). Mutant Norwalk virus VLPs that have point mutations in their capsids which abrogate binding to carbohydrate ligands also do not bind to shellfish tissues (58). Binding of Norwalk virus VLPs to an HBGA A-like carbohydrate has also been found in another oyster species, *Crassostrea virginica* (92). In vitro studies with additional NoV genotypes are needed to determine whether binding to shellfish tissues is genotype specific. With such information, epidemiologic studies of shellfish-associated outbreaks can be designed to address the importance of specific binding of NoV to shellfish tissues.

In bioaccumulation studies, some Norwalk virus particles can be detected in oyster hemophagocytes (58, 85). It is unclear whether the viral nucleic acids (85) or immunoreactive material (58) detected in phagocytes corresponds to particles being degraded and digested or whether particles are able to escape digestion. Internalization of intact viral particles into cells in the intestine or into hemophagocytes may provide another explanation for the poor efficiency of depuration in removing viruses from shellfish.

VIRUS INACTIVATION AND PERSISTENCE

Factors Influencing Virus Inactivation

A number of physical, chemical, and biological factors influence the rate of virus inactivation in the environment (Fig. 2) (6, 81, 82). Exposures to higher temperatures, UV light, lower relative humidity, high pressure, and radiation are physical factors that all contribute to loss of virus infectivity. Chemical factors that increase viral inactivation include exposure to extreme pH values (very acidic or alkaline pH), high salinity, certain enzymes (e.g., proteases and nucleases), ammonia and certain other ions, and a variety of compounds. The presence of bacteria and protozoa introduces external biological factors that can increase virus inactivation (6). The addition of food additives, such as sodium bisulfite or ascorbic acid, may also increase virus inactivation (83). Virus aggregation and the presence of organic matter may protect viruses from inactivation. While the contributions of many of these factors have been studied in a variety of circumstances, there are relatively few studies examining their role in the inactivation of viruses associated with foods.

Effects of Temperature on Virus Inactivation in Foods

Temperature is one of the physical factors best recognized to play a role in virus stability in food. Storage at refrigerator temperatures (4°C) is a common approach to prolonging the fitness of food for consumption, but storage at these temperatures also prolongs virus survival relative to storage at room temperature. For example, Bidawid et al. (5) found approximately a 2-fold reduction in the HAV titer on lettuce stored at 4°C for 12 days compared to a 10,000-fold reduction in titer when the lettuce was stored at room temperature. Virus survival on foods stored at 4°C can be significantly prolonged, as indicated by the detection, for more than 2 months, of infectious poliovirus on celery that had been irrigated with virus-seeded wastewater (95). The level of persistence varies with the vegetable matrix. No decline in infectious poliovirus titer was observed after 2 weeks at 4°C on green onions or fresh raspberries, but a 10-fold reduction was observed after 11.6 days on lettuce, 14.2 days on white cabbage, and 8.4 days on frozen strawberries (54).

Higher temperatures, such as those achieved in cooking or pasteurization, increase the rate of virus inactivation. However, certain constituents or additives in foods may stabilize the virus, protecting it from inactivation. For example, heat inactivation of HAV is less efficient in dairy products containing a greater fractional content of fat (e.g., cream) than in products with less fat (e.g., skim milk) (4), and a higher fat content in ground beef decreases the thermal inactivation of poliovirus (29). Higher sucrose concentrations, such as those used as stabilizers in some fruit-based products, increase HAV resistance to heat inactivation in strawberries (22). On the other hand, higher acidity can increase viral susceptibility to heat inactivation (22).

Effects of Compounds in Foods on Viral Growth

Compounds in foods may inhibit the growth of or inactivate viruses contaminating those foods. Konowalchuk and Speirs (49, 50) conducted a series of experiments examining the effects of a variety of fruit and fruit products on virus growth. Strawberry extracts inhibited poliovirus replication, and the inhibition was attributed to the presence of phenolic compounds in the fruit (50). Phenolic compounds were also thought to be responsible for the inhibitory activity of grape juice, red wines, and white wines against several viruses (coxsackievirus, echovirus, poliovirus, reovirus, and herpes simplex virus) (49). The skin and seeds of grapes, which contain the greatest concentration of phenolic compounds, also had the greatest inhibitory activity. This may also explain the observation that red wines had greater inhibitory activity than white wines, which are produced from fermentation of only the juice (4). The inhibitory activity in grape juice was reversible, suggesting that the virus is not inactivated. Fresh apple juice irreversibly inhibited poliovirus replication, while many other fruit juices (orange, pineapple, grapefruit, and tomato) had no measurable antiviral activity (51). The inhibitory activity of fresh apple juice was lost during storage. This activity was postulated to be due to the presence of tannins in the apple; oxidation of the tannins during storage could explain the loss of the potency of apple juice over time (51). Rotavirus infectivity was stable for 3 days at 4°C in a commercial fruit punch (75). Additional studies are needed to determine the exact mechanisms of action of the inhibitory compounds in these foods.

Effects of Other Environmental Factors on Virus Inactivation

The UV component of sunlight acts as the principal natural germicide in the environment (64, 88). The UV light spectrum can be divided into three segments: UV-A (315 to 400 nm), UV-B (280 to 315 nm), and UV-C (200 to 280 nm). The UV-C portion of the spectrum is the most virucidal, but most of the UV-C rays in sunlight are absorbed by the atmosphere before reaching the Earth's surface. It is therefore estimated that the majority of the virucidal activity in sunlight comes from the UV-B portion of the spectrum (64).

Environmental factors, including the presence of clouds, dust, and pollution, can decrease the penetration of UV-B light through the atmosphere. Double-stranded DNA viruses (e.g., adenoviruses) are more resistant to UV inactivation than are single-stranded RNA viruses (e.g., picornaviruses and caliciviruses), and viruses with larger genomes are more sensitive than those with smaller genomes (33, 87, 88, 91). It is estimated that a full day of sun exposure at latitudes below 37° would produce approximately a 1,000-fold decrease in infectivity for the most UV-sensitive viruses (64). However, limited experimental data are available on the effects of sunlight on the survival of enteric viruses in the environment or on foods.

Field experiments are not able to distinguish which environmental factors are responsible for viral inactivation, but they can show the persistence of viral infectivity. Ward et al. (94) demonstrated that poliovirus could still be recovered from cabbage for up to 5 days after spray irrigation with virus-seeded wastewater, despite mean maximal temperatures ranging from 16 to 22°C. Subsequent experiments showed that poliovirus could be detected, for 4 to 13 days after exposure, on vegetables (celery, spinach, and tomatoes) grown and irrigated with poliovirus-seeded wastewater (95). If the celery and spinach were collected just after irrigation and stored in a refrigerator, virus infectivity was still identifiable for 55 to 76 days (95).

Persistence of Viruses in Shellfish

As noted above, shellfish bioaccumulate enteric viruses which persist under depuration conditions that are sufficient for the shellfish to clear enteric bacteria. This has led to outbreaks of viral disease associated with the consumption of shellfish meeting bacteriological sanitary standards (7, 57, 60, 65, 72, 84). An important question is how long shellfish can retain infectious viruses after bioaccumulation. The presence of virus carriage for an extended period was suggested as early as 1973, when an outbreak of hepatitis A was linked to shellfish consumption (68). The shellfish were thought to have been contaminated following a flood in the spring, and they retained infectious virus for 1 to 2 months before the outbreak. Subsequently, laboratory studies confirmed that infectious HAV could be detected for up to 3 weeks in oyster tissues and RT-PCR methods detected viral RNA for up to 6 weeks (47). In another laboratory experiment, infectious adenovirus was detectable in mussels for approximately 3 weeks and in oysters for up to 7 weeks when maintained in seawater at 4°C (37). Viral infectivity declined more rapidly when the shellfish were maintained at a higher temperature (18°C). As in the earlier study, viral nucleic acids remained detectable for a longer period than did infectivity. VLPs have been used as a surrogate for infectious viruses in field studies and could be detected for several weeks under relaying conditions in an estuary (62).

The duration of virus persistence under natural conditions was examined following a NoV outbreak. Oysters were collected from a harvesting area implicated as the source of shellfish associated with an outbreak, and NoV RNA could be detected approximately 4 weeks after the outbreak was identified (60). Although the oysters met regulatory standards for the presence of *E. coli*, continued exposure to viral contaminants cannot be excluded. NoV infectivity could not be assessed due to the lack of an in vitro culture system.

TREATMENT OF FOODS TO REMOVE OR INACTIVATE VIRUSES

A variety of methods are being used to try to reduce the risk of food-borne transmission of contaminating pathogens. Rinsing food with water is a common practice to remove visible dirt and contaminants, but it is unreliable at removing viruses. For instance, Croci et al. (18) found that HAV persisted on the surface of a variety of fresh produce despite being rinsed in tap water for 5 min. The use of a disinfectant to inactivate contaminating viruses has also been evaluated. Dawson et al. (21) found that the use of chlorinated water (100 ppm of free chlorine) as a wash was not effective at removing the enteric virus surrogate MS2 bacteriophage from the surface of any of the fresh produce tested. Gulati et al. (34) tested the ability of a variety of commercially available disinfectants to inactivate feline calicivirus (a NoV surrogate) applied to a food preparation surface or to strawberries and lettuce. Only peroxyacetic acid and a hydrogen peroxide-containing product effectively decontaminated strawberries and lettuce, but they were effective only at a fourfold-higher concentration than recommended by the manufacturers (34). This approach does not hold much promise for use against NoV due to the poor activity of disinfectants against these viruses (28).

Modified atmospheric packaging (MAP) is a method used to extend the shelf life of fresh produce. With MAP, the produce is packaged while being exposed to elevated carbon dioxide or reduced oxygen concentrations, and these altered gas concentrations inhibit the growth of bacteria that may lead to spoilage. Bidawid et al. (5) examined the effect of different MAP conditions on the survival of HAV contaminating the surface of lettuce. No reduction in the survival of HAV was noted, and under some MAP conditions using high carbon dioxide concentrations, HAV survival increased.

High-hydrostatic-pressure processing (HHP) has emerged as a promising technology for virus inactivation. This method exposes the food to hydrostatic pressures of 200 to 400 MPa for periods of 1 to 5 min. HHP efficiently inactivates enteric viruses suspended in buffer, but the inactivation rates are affected by treatment temperature and virus strain (16, 46). HHP has been applied to shellfish that have bioaccumulated HAV or a murine NoV strain, and greater than 1,000-fold reductions in viral titer were achieved by treatment with ≤400 MPa for 5 min at 5°C (11, 45). Longer treatment times are required to achieve the same level of inactivation of the virus stock when the virus is oyster associated compared to when it is suspended in buffer (45). A potential disadvantage of this method is that changes in the character of the shellfish have been demonstrated in organoleptic studies, and some consumers prefer to eat live oysters (20).

Some foods are treated by marination, and the effect of this process has been evaluated for mussels contaminated with NoV or HAV. The commercial marination process is a two-stage procedure including a preliminary heat treatment (immersion in boiling water or steaming for 3 min) and then marination for several weeks. After 4 weeks of marination, the infectious titer of HAV had decreased approximately 50-fold and human NoV RNA was still detected by real-time RT-PCR (38). These data suggest that marination alone is not sufficient to inactivate enteric viral pathogens in shellfish.

Gamma irradiation of food is being explored as a potential means of inactivating enteric pathogens that contaminate the food (55). However, there is limited information evaluating the efficacy of this technology on viral pathogens. The dosages (400 Gy) needed to inactivate two NoV surrogates (feline calicivirus and canine calicivirus) at least 1,000-fold are well within the ranges being evaluated in other food irradiation studies (24, 55).

Cooking is the most reliable manner of inactivating viruses in foods, but there are numerous examples where "cooked foods" still contained infectious virus and were able to transmit disease (Tables 1 and 2). Early studies showed that 7 to 13% of poliovirus in oysters survived a variety of home-cooking procedures (steaming, stewing, frying, and baking) (25). Steaming bivalve mollusks for 5 min or less does not achieve a sufficient temperature in the shellfish meat for viral inactivation (48). In mussels, HAV and human rotavirus could still be detected 5 min after the opening of the valves by steaming, showing a reduction in the original titer below 3 log units, whereas poliovirus was no longer detectable (1). HAV-contaminated cockles needed to be immersed for at least 1 min in boiling water to achieve HAV inactivation (69), and viable HAV persisted in contaminated mussels after boiling for 37 s, although the HAV was completely inactivated after boiling for 3 min (39). Differences observed in virus reduction may be due in part to the method of virus inoculation, i.e., bioaccumulation versus artificial seeding, suggesting that virus localization may play an important protective role. Unfortunately, many of the

Table 1 Effects of different physical factors on virus inactivation in shellfish

Type of treatment	Recipe	Virus	Contamination	Persistence of virus	Reference
Heat	Stewed oysters in milk	Poliovirus	Seeded	10% survival after 8 min	25
	Fried oysters in oil at 177°C	Poliovirus	Seeded	10% survival after 8 min	25
	Baked oysters in the oven at 121°C	Poliovirus	Seeded	13% survival after 20 min	25
	Boiled cockles	Poliovirus	Bioaccumulated	No survival after 3 min	69
		HAV		No survival after 2 min	
	Boiled mussels	HAV	Seeded	No survival after 3 min for HAV	39
		NoV		100% persistence for NoV RNA	
	Steamed mussels	HAV	Bioaccumulated	0.1% survival after 5 min	1
		Rotavirus		0.1% survival after 5 min	
	Steamed mussels	HAV	Seeded	No survival after 3 min for HAV	39
		NoV		100% persistence for NoV RNA	
	Steamed oysters	Poliovirus	Bioaccumulated	7% survival after 30 min	25
	Mussels in hors d'oeuvres	HAV	Bioaccumulated	Infectious viruses detected after 9 min[a]	17
	Mussels au gratin	HAV	Bioaccumulated	Infectious viruses detected after 5 min	17
	Mussels in tomato sauce	HAV	Bioaccumulated	No survival after 8 min	17
Cold	Oysters kept at 5°C	Poliovirus	Bioaccumulated	13% survival after 1 mo	25
	Frozen oysters (−17.5°C)	Poliovirus	Bioaccumulated	10% survival after 12 wk	25
Marinade	Steamed mussels and marination (pH 3. 75)	HAV	Seeded	3% survival after 4 wk for HAV	38
		NoV		100% persistence for NoV RNA	
High pressure	Oysters	Murine NoV	Bioaccumulated	No survival after 5 min at 5°C at 325 MPa	45
	Oysters	HAV	Bioaccumulated	0.1% survival after 1 min at 400 MPa	11

[a]Infectious virus was present in numbers too small for quantification.

Table 2 Selected outbreaks linked to prepared foods

Food	Virus	Cooking/food preparation	No. of cases	Reference
Shellfish				
Cockles	HAV	Steaming for 1–2 min and boiling for 4 min	132	76
Coquina clams	HAV	Frozen and paella cooking	183	7
Oysters	NoV	Kilpatrick or mornay	>2,000	70
Oysters	NoV	Grilled, steamed, stewed, or fried	131	65
Oysters	NoV	Frozen and served thawed	213	72
Clams, oysters	NoV, AV[a]	Clam soup and cooked oysters from frozen stock	26	84
Mussels	NoV	Cooked	89	78
Other food				
Pig liver	HEV	Grilled	10	99
Iceberg lettuce	HAV	Washed	202	80
Rucola lettuce	HAV	Ready to eat	80	73
Strawberries	HAV	Incorporated in a shortcake	128	40
Green onions	HAV	Melted cheese sauce, rice	43	23
Orange juice	HAV	Manufacturing process	351	30
Raspberries	NoV	Frozen dressing	106	77
Raspberries	NoV	Cake	300	31
Raspberries	NoV	Cake	30	59

[a]AV, astrovirus.

cooking methods that are sufficient to inactivate contaminating enteric viruses have an adverse effect on the palatability of the shellfish for consumption.

CONSEQUENCES OF VIRUS PERSISTENCE IN FOODS

The inability to remove or inactivate enteric viruses from contaminated foodstuffs leads inevitably to human disease in some susceptible consumers. There are numerous examples of outbreaks of gastroenteritis or hepatitis associated with the consumption of raw or partially cooked foodstuffs (Table 2). Some shellfish-related outbreaks clearly show that the cooking procedures used did not efficiently inactivate contaminating viruses (Table 2). For example, fried oysters have been responsible for illness, as have boiled cockles and clams in paella. This resistance to heating is also seen in other foods, such as grilled liver (99). Processed foods, such as packaged lettuce, orange juice, and raspberry cake, have been implicated in outbreaks. Contamination of food during preparation may also occur. It is to be expected that such outbreaks will continue to occur as long as there are opportunities for viral contamination of foodstuffs and these foods are consumed without undergoing treatments that effectively inactivate the contaminating viruses.

CONCLUSIONS AND SUMMARY

Enteric viruses bind to food matrices by a variety of mechanisms, including ionic and hydrophobic interactions, van der Waals forces, interaction with specific ligands (e.g., receptors), and uptake into plants. Removal or inactivation of viruses contaminating foods has proven to be difficult; ineffective removal or inactivation can lead to food-borne outbreaks of gastroenteritis or hepatitis.

The role of the food matrix in virus persistence is difficult to estimate, since few data addressing this issue are available. To collect more information, it will be important to work on improving methods of virus detection, since it is still difficult to detect a virus in a complex matrix due to low levels of virus contamination, the presence of compounds toxic in cell culture, and the presence of inhibitors of molecular detection assays. Without sensitive and efficient methods of detection, it will be difficult to obtain information on virus persistence and localization, which in turn may provide insights that lead to improved detection assays. For example, the recognition that most of the enteric viruses contaminating shellfish are located in pancreatic tissues led to a more than 10-fold increase in the sensitivity of virus detection by eliminating shellfish tissues that contained inhibitors of PCR. The identification of HAV inside green onion tissues demonstrates that simple washing of the surface of a food may be insufficient to identify viruses responsible for outbreaks of disease; instead, all plant tissues may have to be analyzed. A better understanding of the interaction between pathogenic enteric viruses and different food matrices should lead to enhanced measures to remove or inactivate these viruses and ultimately to improved food safety.

ACKNOWLEDGMENTS

We thank M. Pommepuy and P. Bodennes for preparation of the figures.

REFERENCES

1. **Abad, F. X., R. M. Pinto, R. Gajardo, and A. Bosch.** 1997. Viruses in mussels: public health implications and depuration. *J. Food Prot.* **60:**677–681.

2. **Abad, F. X., C. Villena, S. Guix, S. Caballero, R . M. Pinto, and A. Bosch.** 2001. Potential role of fomites in the vehicular transmission of human astroviruses. *Appl. Environ. Microbiol.* **67:**3904–3907.

3. Amon, J. J., R. Devasia, G. Xia, O. V. Nainan, S. Hall, B. Lawson, J. S. Wolthuis, P. D. Macdonald, C. W. Shepard, I. T. Williams, G. L. Armstrong, J. A. Gabel, P. Erwin, L. Sheeler, W. Kuhnert, P. Patel, G. Vaughan, A. Weltman, A. S. Craig, B. P. Bell, and A. Fiore. 2005. Molecular epidemiology of foodborne hepatitis A outbreaks in the United States, 2003. *J. Infect. Dis.* **192:**1323–1330.

4. Bidawid S., J. M. Farber, S. A. Sattar, and S. Hayward. 2000. Heat inactivation of hepatitis A virus in dairy foods. *J. Food Prot.* **63:**522–528.

5. Bidawid, S., J. M. Farber, and S. A. Sattar. 2001. Survival of hepatitis A virus on modified atmosphere-packaged (MAP) lettuce. *Food Microbiol.* **18:**95–102.

6. Bosch, A., R. M. Pinto, and F. X. Abad. 2006. Survival and transport of enteric viruses in the environment, p. 151–187. *In* S. M. Goyal (ed.), *Viruses in Foods*. Springer, New York, NY.

7. Bosch, A., G. Sanchez, F. Le Guyader, H. Vanaclocha, L. Haugarreau, and R. M. Pinto. 2001. Human enteric viruses in coquina clams associated with a large hepatitis A outbreak. *Water Sci. Technol.* **43:**61–65.

8. Bowen, A., A. Fry, G. Richards, and L. Beauchat. 2006. Infections associated with cantaloupe consumption: a public health concern. *Epidemiol. Infect.* **134:**675–685.

9. Burkhardt, W., and K. R. Calci. 2000. Selective accumulation may account for shellfish-associated viral illness. *Appl. Environ. Microbiol.* **66:**1375–1378.

10. Butot, S., T. Putallaz, and G. Sanchez. 2007. Procedure for rapid concentration and detection of enteric viruses from berries and vegetables. *Appl. Environ. Microbiol.* **73:**186–192.

11. Calci, K. R., G. K. Meade, R. C. Tezloff, and D. H. Kingsley. 2005. High-pressure inactivation of hepatitis A virus within oysters. *Appl. Environ. Microbiol.* **71:**339–343.

12. Carter, M. J. 2005. Enterically infecting viruses: pathogenicity, transmission and significance for food and waterborne infection. *J. Appl. Microbiol.* **98:**1354–1380.

13. Centers for Disease Control and Prevention. 2000. Foodborne outbreak of group A rotavirus gastroenteritis among college students—District of Columbia, March–April 2000. *Morb. Mortal. Wkly. Rep.* **49:**1131–1133.

14. Chancellor, D. D., S. Tyagi, M. C. Bazaco, S. Bacvinskas, M. B. Chancellor, V. M. Dato, and F. de Miguel. 2006. Green onions: potential mechanism for hepatitis A contamination. *J. Food Prot.* **69:**1468–1472.

15. Chattopadhyay, S., and R. W. Puls. 1999. Adsorption of bacteriophages on clay minerals. *Environ. Sci. Technol.* **33:**3609–3614.

16. Chen, H., D. G. Hoover, and D. H. Kinsgley. 2005. Temperature and treatment time influence high hydrostatic pressure inactivation of feline calicivirus, a norovirus surrogate. *J. Food Prot.* **68:**2389–2394.

17. Croci, L., D. De Medici, S. Di Pasquale, and L. Toti. 2005. Resistance of hepatitis A virus in mussels subjected to different domestic cookings. *Int. J. Food Microbiol.* **105:**139–144.

18. Croci, L., D. De Medici, C. Scalfaro, A. Fiore, and L. Toti. 2002. The survival of hepatitis A virus in fresh produce. *Int. J. Food Microbiol.* **73:**29–34.

19. Cromeans, T. L., and M. D. Sobsey. 2004. Environmentally transmissible enteric hepatitis viruses: A and E, p. 2267–2274. *In* R. K. Robinson (ed.), *Encyclopedia of Food Microbiology*. Elsevier, Oxford, United Kingdom.

20. Cruz-Romero, M., M. Smiddy, C. Hill, J. P. Kerry, and A. L. Kelly. 2004. Effects of high pressure treatment on physicochemical characteristics of fresh oysters (*Crassostrea gigas*). *Innov. Food Sci. Emerg. Technol.* **5**:161–169.

21. Dawson, D. J., A. Paish, L. M. Stafell, I. J. Seymour, and H. Appleton. 2005. Survival of viruses on fresh produce, using MS2 as a surrogate for norovirus. *J. Appl. Microbiol.* **98**:203–209.

22. Deboosere, N., O. Legeay, Y. Caudrelier, and M. Lange. 2004. Modelling effect of physical and chemical parameters on heat inactivation kinetics of hepatitis A virus in a fruit model system. *Int. J. Food Microbiol.* **93**:73–85.

23. Dentinger, C. M., W. A. Bower, O. V. Nainan, S. M. Cotter, G. Myers, L. M. Dubusky, S. Fowler, E. D. Salehi, and B. P. Bell. 2001. An outbreak of hepatitis A associated with green onions. *J. Infect. Dis.* **183**:1273–1276.

24. de Roda Husman, A. M., P. Bijkerk, W. Lodder, H Van Den Berg, W. Pribil, A. Cabaj, P. Gehringer, R. Sommer, and E. Duizer. 2004. Calicivirus inactivation by nonionizing (253.7-nanometer-wavelength [UV]) and ionizing (gamma) radiation. *Appl. Environ. Microbiol.* **70**:5089–5093.

25. Di Girolamo, R., J. Liston, and J. Matches. 1970. Survival of virus in chilled, frozen, and processed oysters. *Appl. Environ. Microbiol.* **20**:58–63.

26. Di Girolamo, R., J. Liston, and J. Matches. 1977. Ionic binding, the mechanism of viral uptake by shellfish mucus. *Appl. Environ. Microbiol.* **33**:19–25.

27. Dubois, E., C. Agier, O. Traore, C. Hennechart, G. Merle, C. Cruciere, and H. Laveran. 2002. Modified concentration method for the detection of enteric viruses on fruits and vegetables by reverse transcriptase-polymerase chain reaction or cell culture. *J. Food Prot.* **65**:1962–1969.

28. Duizer, E., P. Bijkerk, B. Rockx, A. de Groot, F. Twisk, and M. Koopmans. 2004. Inactivation of caliciviruses. *Appl. Environ. Microbiol.* **70**:4538–4543.

29. Filppi, J. A., and G. J. Banwart. 1974. Effect of the fat content of ground beef on the heat inactivation of poliovirus. *J. Food Sci.* **39**:865–868.

30. Franck, C., J. Walter, M. Muehlen, A. Jansen, U. van Treeck, A. M. Hauri, I. Zoellner, M. Rakha, M. Hoehne, O. Hamouda, E. Scheier, and K. Stark. 2007. Major outbeak of hepatitis A associated with orange juice among tourists, Egypt, 2004. *Emerg. Infect. Dis.* **13**:156–158.

31. Gaulin, C. D., D. Ramsay, P. Cardinal, and M-A D'Halevyn. 1999. Epidemie de gastro-enterite d'origine virale associée à la consommation de framboises importées. *Can. J. Public Health* **90**:37–40.

32. Gerba, C. P. 1984. Applied and theoretical aspects of virus adsorption to surfaces. *Adv. Appl. Microbiol.* **30**:133–168.

33. Gerba, C. P., D. M. Gramos, and N. Nwachuku. 2002. Comparative inactivation of enteroviruses and adenovirus 2 by UV light. *Appl. Environ. Microbiol.* **68**:5167–5169.

34. Gulati, B. R., P. B. Allwood, C. W. Hedberg, and S. M. Goyal. 2001. Efficacy of commonly used disinfectants for the inactivation of calicivirus on strawberry, lettuce and a food-contact surface. *J. Food Prot.* **64**:1430–1434.

35. Guo, X., M. W. van Iersel, J. Chen, R. E. Brackett, and L. R. Beuchat. 2002. Evidence of association of salmonellae with tomato plants grown hydroponically in inoculated nutrient solution. *Appl. Environ. Microbiol.* **68**:3639–3643.

36. **Hernandez, F., R. Monge, C. Jimenez, and L. Taylor.** 1997. Rotavirus and hepatitis A virus in market lettuce (*Latuca sativa*) in Costa Rica. *Int. J. Food Microbiol.* **37:**221–223.

37. **Hernroth, B., and A. Allard.** 2007. The persistence of infectious adenovirus (type 35) in mussels (*Mytilus edulis*) and oysters (*Ostrea edulis*). *Int. J. Food Microbiol.* **113:**296–302.

38. **Hewitt, J., and G. E. Greening.** 2004. Survival and persistence of norovirus, hepatitis A virus, and feline calicivirus in marinated mussels. *J. Food Prot.* **67:**1743–1750.

39. **Hewitt, J., and G. E. Greening.** 2006. Effect of heat treatment on hepatitis A virus and norovirus in New Zealand greenshell mussels (*Perna canaliculus*) by quantitative real-time reverse transcription PCR and cell culture. *J. Food Prot.* **69:**2217–2223.

40. **Hutin, Y. F. H., V. Pool, E. H. Cramer, O. V. Nainan, J. Weth, I. T. Williams, S. T. Goldstein, K. F. Gensheimer, B. P. Bell, C. N. Shapiro, M. J. Alter, and H. S. Margolis.** 1999. A multistate, foodborne outbreak of hepatitis A. *N. Engl. J. Med.* **340:**595–602.

41. **Hutson, A. M., F. Airaud, J. Le Pendu, M. K. Estes, and R. L. Atmar.** 2005. Norwalk virus infection associates with secretor status genotyped from sera. *J. Med. Virol.* **77:**116–120.

42. **Jin, Y., and M. Flury.** 2002. Fate and transport of viruses in porous media. *Adv. Agron.* **77:**39–102.

43. **Katzenelson, E., and D. Mills.** 1984. Contamination of vegetables with animal viruses via the roots. *Monogr. Virol.* **15:**216–220.

44. **Kerbo, N., I. Donchenko, K. Kutsar, and V. Vasilenko.** 2005. Tickborne encephalitis outbreak in Estonia linked to raw goat milk, May–June 2005. *Eur. Surveill.* **10:**E050623.2.

45. **Kingsley, D. H., D. R. Holliman, K. R. Calci, H. Chen, and G. J. Flick.** 2007. Inactivation of a norovirus by high-pressure processing. *Appl. Environ. Microbiol.* **73:**581–585.

46. **Kingsley, D. H., D. G. Hoover, E. Papafragkou, and G. P. Richards.** 2002. Inactivation of hepatitis A virus and a calicivirus by high hydrostatic pressure. *J. Food Prot.* **65:**1605–1609.

47. **Kingsley, D. H., and G. P. Richards.** 2003. Persistence of hepatitis A virus in oysters. *J. Food Prot.* **66:**331–334.

48. **Koff, R. S., and H. S. Sears.** 1967. Internal temperature of steamed clams. *N. Engl. J. Med.* **276:**737–739.

49. **Konowalchuk, J., and J. I. Speirs.** 1976. Virus inactivation by grapes and wines. *Appl. Environ. Microbiol.* **32:**757–763.

50. **Konowalchuk, J., and J. I. Speirs.** 1978. Antiviral effect of commercial juices and beverages. *Appl. Environ. Microbiol.* **35:**1219–1220.

51. **Konowalchuk, J., and J. I. Speirs.** 1978. Antiviral effect of apple beverages. *Appl. Environ. Microbiol.* **36:**798–801.

52. **Koopmans, M., and E. Duizer.** 2004. Foodborne viruses: an emerging problem. *Int. J. Food Microbiol.* **90:**23–41.

53. **Kukavica-Ibrilj, I., A. Darveau, J. Jean, and I. Fliss.** 2004. Hepatitis A virus attachment to agri-food surfaces using immunological and thermodynamic assays. *J. Appl. Microbiol.* **97:**923–934.

54. **Kurdziel, A. S., N. Wilkinson, S. Langton, and N. Cook.** 2001. Survival of poliovirus on soft fruits and salad vegetables. *J. Food Prot.* **64:**706–709.

55. Lee, N. Y., C. Jo, D. H. Shin, W. G. Kim, and M. W. Byun. 2006. Effect of gamma-irradiation on pathogens inoculated into ready-to-use vegetables. *Food Microbiol.* **23**:649–656.

56. Leggitt, P. R., and L.-A. Jaykus. 2000. Detection methods for human enteric viruses in representative foods. *J. Food Prot.* **63**:1738–1744.

57. Le Guyader, F. S., F. Bon, D. De Medici, S. Parnaudeau, A. Bertome, S. Crudeli, A. Doyle, M. Zidane, E. Suffredini, E. Kohli, F. Maddalo, M. Monini, A. Gallay, M. Pommepuy, P. Pothier, and F. Ruggeri. 2006. Detection of noroviruses in an international gastroenteritis outbreak linked to oyster consumption. *J. Clin. Microbiol.* **44**:3878–3882.

58. Le Guyader, F. S., F. Loisy, R. L. Atmar, A. M. Hutson, M. K. Estes, N. Ruvoen-Clouet, M. Pommepuy, and J. Le Pendu. 2006. Norwalk virus specific binding to oyster digestive tissues. *Emerg. Infect. Dis.* **12**:931–936.

59. Le Guyader, F. S., C. Mittelholzer, L. Haugarreau, K.-O. Hedlund, R. Alsterlund, M. Pommepuy, and L. Svensson. 2004. Detection of noroviruses in raspberries associated with a gastroenteritis outbreak. *Int. J. Food Microbiol.* **97**:179–186.

60. Le Guyader, F. S., F. H. Neill, E. Dubois, F. Bon, F. Loisy, E. Kohli, M. Pommepuy, and R. L. Atmar. 2003. A semi-quantitative approach to estimate Norwalk-like virus contamination of oysters implicated in an outbreak. *Int. J. Food Microbiol.* **87**:107–112.

61. Li, T. C., K. Chijiwa, N. Sera, T. Ishibashi, Y. Etoh, Y. Shinohara, Y. Kurata, M. Ishida, S. Sakamoto, N. Takeda, and T. Miyamura. 2005. Hepatitis E virus transmission from wild boar meat. *Emerg. Infect. Dis.* **11**:1958–1960.

62. Loisy, F., R. L. Atmar, J.-C. Le Saux, J. Cohen, M.-P. Caprais, M. Pommepuy, and F. S. Le Guyader. 2005. Use of rotavirus virus-like particles as surrogates to evaluate virus persistence in shellfish. *Appl. Environ. Microbiol.* **71**:6049–6053.

63. Luby, S. P., M. Rahman, M. J. Hossain, L. S. Blum, M. M. Husain, E. Gurley, R. Khan, B.-N. Ahmed, S. Rahman, N. Nahar, E. Kenah, J. A. Comer, and T. G. Ksiazek. 2006. Foodborne transmission of Nipah virus, Bangladesh. *Emerg. Infect. Dis.* **12**:1888–1894.

64. Lytle, C. D., and J.-L. Sagripanti. 2005. Predicted inactivation of viruses of relevance to biodefense by solar radiation. *J. Virol.* **79**:14244–14252.

65. MacDonnel, S., K. B. Kirkland, W. G. Hlady, C. Aristeguita, R. S. Hopkins, S. S. Monroe, and R. I. Glass. 1997. Failure of cooking to prevent shellfish-associated viral gastroenteritis. *Arch. Intern. Med.* **157**:111–116.

66. Mazur, B., and W. Paciorkiewicz. 1973. Dissemination of enteroviruses in the human environment. I. Presence of poliovirus in various parts of vegetable plants grown on infected soil. *Med. Dosw. Mikrobiol.* **25**:93–98.

67. Mead, P. S., L. Slutsker, V. Dietz, L. F. McCaig, J. S. Bresee, C. Shapiro, P. M. Griffin, and R. V. Tauxe. 1999. Food-related illness and death in the United States. *Emerg. Infect. Dis.* **5**:607–625.

68. Metcalf, T. G. 1982. Viruses in shellfish growing waters. *Environ. Int.* **7**:21–27.

69. Millard, J., H. Appleton, and J. V. Parry. 1987. Studies on heat inactivation of hepatitis A virus with special reference to shellfish. *Epidemiol. Infect.* **98**:397–414.

70. Murphy, A. M., G. S. Grohmann, P. J. Christopher, W. A. Lopez, G. R. Davey, and R. H. Millson. 1979. An Australia-wide outbreak of gastroenteritis from oysters caused by Norwalk virus. *Med. J. Aust.* **2**:329–333.

71. Murphy, W. H., and J. T. Syverton. 1958. Absorption and translocation of mammalian viruses by plants. II. Recovery and distribution of viruses in plants. *Virology* 6:623–636.

72. Ng, T. L., P. P. Chan, T. H. Phua, J. P. Loh, R. Yip, C. Wong, C. W. Liaw, B. H. Tan, K. T. Chiew, S. B. Chua, S. Lim, P. L. Ooi, S. K. Chew, and K. T. Goh. 2005. Oyster-associated outbreaks of norovirus gastroenteritis in Singapore. *J. Infect.* 51:413–418.

73. Nygard, K., Y. Andersson, P. Lindkvist, C. Ancker, I. Asteberg, E. Dannetun, R. Eitrem, L. Hellstrom, M. Insulander, L. Skedebrant, K. Stenqvist, and J. Giesecke. 2001. Imported rocket salad partly responsible for increased incidence of hepatitis A cases in Sweden, 2000–2001. *Eur. Surveill.* 6:151–153.

74. Oishi, I., K. Yamazaki, T. Kimoto, Y. Minekawa, E. Utagawa, S. Yamazaki, S. Inouye, G. S. Grohmann, S. S. Monroe, S. E. Stine, C. Carcamo, T. Ando, and R. I. Glass. 1994. A large outbreak of acute gastroenteritis associated with astrovirus among students and teachers in Osaka, Japan. *J. Infect. Dis.* 170:439–443.

75. O'Mahony, M., O. Donoghue, J. G. Morgan, and C. Hill. 2000. Rotavirus survival and stability in foods as determined by an optimised plaque assay procedure. *Int. J. Food Microbiol.* 61:177–185.

76. O'Mahony, M., C. D. Gooch, D. A. Smyth, A. J. Trussel, C. L. R. Bartlett, and N. D. Noah. 1983. Epidemic hepatitis A from cockles. *Lancet* i:518–520.

77. Ponka, A., L. Maunula, C. H. Von Bonsdorff, and O. Lyytikainen. 1999. An outbreak of calicivirus associated with consumption of frozen raspberries. *Epidemiol. Infect.* 123:469–474.

78. Prato, R., P. L. Lopalco, M. Chironna, G. Barbuti, C. Germinario, and M. Quarto. 2004. Norovirus gastroenteritis general outbreak associated with raw shellfish consumption in South Italy. *BMC Infect. Dis.* 4:1–6.

79. Richards, G. P. 1988. Microbial purification of shellfish: a review of depuration and relaying. *J. Food Prot.* 51:218–251.

80. Rosenblum, L. S., I. R. Mirkin, D. T. Allen, S. Safford, and S. C. Hadler. 1990. A multifocal outbreak of hepatitis A traced to commercially distributed lettuce. *Am. J. Public Health* 80:1075–1079.

81. Rzezutka, A., and N. Cook. 2004. Survival of human enteric viruses in the environment and food. *FEMS Microbiol. Rev.* 28:441–453.

82. Salo R. J., and D. O. Cliver. 1976. Effect of acid pH, salts, and temperature on the infectivity and physical integrity of enteroviruses. *Arch. Virol.* 52:269–282.

83. Salo, R. J., and D. O. Cliver. 1978. Inactivation of enteroviruses by ascorbic acid and sodium bisulfite. *Appl. Environ. Microbiol.* 36:68–75.

84. Sasaki, Y., A. Kai, Y. Hayashi, T. Shinkai, Y. Nogushi, M. Hasegawa, K. Sadamasu, K. Mori, Y. Tabei, M. Nagashima, S. Morozumi, and T. Yamamoto. 2006. Multiple viral infections and genomic divergence among noroviruses during an outbreak of acute gastroenteritis. *J. Clin. Microbiol.* 44:790–797.

85. Schwab, K. J., F. H. Neill, M. K. Estes, T. G. Metcalf, and R. L. Atmar. 1998. Distribution of Norwalk virus within shellfish following bioaccumulation and subsequent depuration by detection using RT-PCR. *J. Food Prot.* 61:1674–1680.

86. Seymour, I. J., and H. Appleton. 2001. Foodborne viruses and fresh produce. *J. Appl. Microbiol.* 91:759–773.

87. Simonet, J., and C. Gantzer. 2006. Inactivation of poliovirus 1 and F-specific RNA phages and degradation of their genomes by UV irradiation at 254 nanometers. *Appl. Environ. Microbiol.* **72**:7671–7677.

88. Sinton, L.W., R. K. Finlay, and P. A. Lynch. 1999. Sunlight inactivation of fecal bacteriophages and bacteria in sewage-polluted seawater. *Appl. Environ. Microbiol.* **65**:3605–3613.

89. Taku, A., B. R. Gulati, P. B. Allwood, K. Palazzi, C. W. Hedberg, and S. M. Goyal. 2002. Concentration and detection of caliciviruses from food contact surfaces. *J. Food Prot.* **65**:999–1004.

90. Tan, M., and X. Jiang. 2005. Norovirus and its histo-blood group antigen receptors: an answer to a historical puzzle. *Trends Microbiol.* **13**:285–293.

91. Thurston-Enriquez, J. A., C. N. Haas, J. Jacangelo, K. Riley, and C. P. Gerba. 2003. Inactivation of feline calicivirus and adenovirus type 40 by UV radiation. *Appl. Environ. Microbiol.* **69**:577–582.

92. Tian, P., A. H. Bates, H. M. Jensen, and R. Mandrell. 2006. Norovirus binds to blood group A-like antigens in oyster gastrointestinal cells. *Lett. Appl. Microbiol.* **43**:645–651.

93. Vega, E., J. Smith, J. Garland, A. Matos, and S. D. Pillaii. 2005. Variability of virus attachment patterns to butterhead lettuce. *J. Food Prot.* **68**:2112–2117.

94. Ward, B. K., C. M. Chenoweth, and L. G. Irving. 1982. Recovery of viruses from vegetable surfaces. *Appl. Environ. Microbiol.* **44**:1389–1394.

95. Ward, B. K., and L. G. Irving. 1987. Virus survival on vegetables spray-irrigated with wastewater. *Water Res.* **21**:57–63.

96. Wheeler, C., T. M. Vogt, G. L. Armstrong, G. Vaughan, A. Weltman, O. V. Nainan, V. Dato, G. Xia, K. Waller, J. Amon, T. M. Lee, A. Highbaugh-Battle, C. Hembree, S. Evenson, M. A. Ruta, I. T. Williams, A. E. Fiore, and B. P. Bell. 2005. An outbreak of hepatitis A associated with green onions. *N. Engl. J. Med.* **353**:890–897.

97. Widdowson, M.-A., A. Sulka, S . N. Bulens, R. Beard, S. S. Chaves, R. Hammond, E. D. P. Salehi, E. Swanson, J. Totaro, R. Woron, P. S. Mead, J. S. Bresee, S. S. Monroe, and R. I. Glass. 2005. Norovirus and foodborne disease, United States, 1991–2000. *Emerg. Infect. Dis.* **11**:95–102.

98. Yamashita, T., M. Sugiyama, H. Tsuzuki, K. Sakae, Y. Suzuki, and Y. Miyazaki. 2000. Application of a reverse transcription-PCR for identification and differentiation of Aichi virus, a new member of the picornavirus family associated with gastroenteritis in humans. *J. Clin. Microbiol.* **38**:2955–2961.

99. Yazaki, Y., H. Mizuo, M. Takahashi, T. Nishizawa, N. Sasaki, Y. Gotanda, and H. Okamoto. 2003. Sporadic acute of fulminant hepatitis E in Hokkaido, Japan, may be food-borne, as suggested by the presence of hepatitis E virus in pig liver as food. *J. Gen. Virol.* **84**:2351–2357.

100. Zheng, D.-P., T. Ando, R. L. Fankhauser, R. S. Beard, R. Glass, and S. S. Monroe. 2006. Norovirus classification and proposed strain nomenclature. *Virology* **346**:312–323.

Food-Borne Viruses: Progress and Challenges
Edited by Marion P. G. Koopmans, Dean O. Cliver, and Albert Bosch
© 2008 ASM Press, Washington, DC

Use of the Codex Risk Analysis Framework To Reduce Risks Associated with Viruses in Food

9

Jaap Jansen

Food virology has reached the stage where scientists are increasingly approached to provide scientific guidance for the possible control of food-borne disease (risk management). The international organizations the World Health Organization (WHO) and the Food and Agriculture Organization (FAO) of the United Nations involved in ensuring effective control of food-borne illness do so through a complex, science-based process. There is an increasing demand for guidance on food-borne viral infections, which are now moving closer to the top of the agenda of these organizations (17). In this chapter, we provide an overview of the organizations involved, the methods used, and the difficulties faced when starting a microbial risk assessment procedure for viruses in food. The potential of the Codex risk analysis framework to reduce risks associated with viruses in food is discussed using a bacterial example.

During the past ~15 years, risk analysis has emerged as a structured approach to improving our food control systems; it is intended to eventually result in the production of safer food and reduce the numbers of food-borne illnesses. Risk analysis is a process consisting of three components: risk assessment, risk management, and risk communication. It is intended to facilitate domestic and international trade in food by promoting harmonized, transparent standards for food and identifying data gaps and information needs. In risk analysis, the entire food production chain is considered in an attempt to produce safer food. Tools for the implementation of microbiological risk management (MRM) (7), such as risk profile, and metrics to quantify risks and the efficacy of control measures have recently become available.

JAAP JANSEN, 614 Rue Villard, 01220 Divonne les Bains, France.

Risk assessment is the first component of risk analysis. It involves the systematic collection of scientific data, including detailed knowledge of the agents and their genetic makeup, growth, survival, inactivation, and health impact. The science also involves knowledge of food production, processing, and preservation, as well as how microorganisms can help and harm humans in these contexts. The risk assessment framework was developed to help organize relevant information; to clarify interactions between microorganisms, foods, and human illness in more detail; and to allow quantification of the risk to human health from specific microorganisms in specific foods. The approach was also developed to allow a comparative evaluation of different possible control options with a view to devising evidence-based interventions. Thus, microbiological risk assessment is currently considered to be the tool of choice when deciding how to control food-borne diseases (Fig. 1).

The remainder of this chapter considers how the Codex Committee for Food Hygiene (CCFH) applies the Codex risk analysis framework to the area of microbial safety, encompassing microbiological risk assessment and management, and how it can be applied in the near future to reduce the burden of food-borne virus disease.

CODEX ALIMENTARIUS

The Codex Alimentarius ("the food code") is a collection of standards, codes of practice, guidelines, and other recommendations (18). Some deal with detailed requirements related to a food or group of foods; others deal with the operation and management of production processes or the operation of government regulatory systems for food safety and consumer protection. United Nations Resolution 39/248 (1985), which adopts guidelines for the use of consumer protection policies, says, "When formulating national policies with regard to food, Governments should . . . as far as possible, adopt standards from the . . . Codex Alimentarius."

Codex and International Food Trade

A principal concern of national governments has always been that food imported from other countries should not jeopardize the health of consumers or pose a threat to the health and safety of their animal and plant populations. Consequently, governments of importing countries have introduced mandatory laws and regulations to eliminate or minimize such threats. In the area of food, animal, and plant control, these measures could be conducive to the creation of barriers to intercountry food trade. The founders of the Codex Alimentarius Commission felt that, if all countries harmonized their

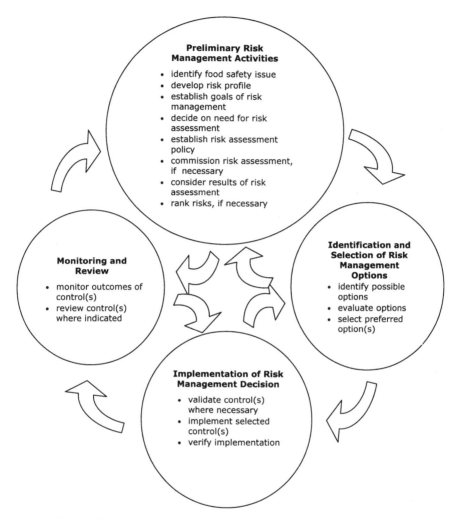

Figure 1 Generic framework for microbiological risk management. (Reprinted from reference 15 with permission.)

food laws by adopting internationally agreed standards, such issues would be resolved.

To prevent the creation of such barriers, the Agreements that established the World Trade Organization (WTO) in 1994 included the SPS (19) and TBT (20) Agreements, specifying what countries can and cannot do in controlling food-related illness. For example, Article 2.2 of the SPS Agreement states, "Members shall ensure that any sanitary and phytosanitary measure is applied only to the extent necessary to protect human, animal or plant life or health, is based on scientific principles and is not maintained without sufficient scientific evidence . . ." The interest in Codex activities has been strongly stimulated by these agreements and the recognition of Codex standards, guidelines, and recommendations. Consequently, attendance at Codex meetings has markedly increased since 1994. Codex standards have become an integral part of the legal framework within which international trade is being facilitated through harmonization. The Codex system has become a platform for countries to formulate food standards and a global reference point for consumers, food producers and processors, national food control agencies, and the international food trade.

Standards, Codes of Practice, and Guidelines

Codex standards usually relate to product characteristics appropriate to the commodity. Maximum residue limits (MRLs) for residues of pesticides or veterinary drugs in foods are examples. There are Codex general standards for food additives, contaminants, and toxins in foods that contain both general and commodity-specific provisions. Codex methods of analysis and sampling are intended to guide both training in and application of sampling and analytical methods. Codex codes of practice define the production, processing, manufacturing, transport, and storage practices that are considered essential to ensure the safety and suitability of food for consumption for individual foods or groups of foods. Hygiene practices are laid down in codes of hygienic practice. For food hygiene, the basic text is the *Codex General Principles of Food Hygiene* (10), which introduces the use of the hazard analysis critical control point (HACCP) food safety management system, the establishment and application of microbiological criteria for foods, and the guidelines for the conduct of microbiological risk assessment.

Codex guidelines fall into two categories: principles that set out policy in certain key areas, and guidelines for the interpretation of these principles. Examples are guidelines for food labeling, especially claims made on the label, including nutrition and health claims; guidelines for production, marketing, and labeling of organic foods; guidelines for Food Import and Export Inspection and Certification; and guidelines for the safety assessments of foods from DNA-modified plants and microorganisms.

Commodity standards are by far the largest group of specific Codex standards, covering commodities such as cereals, fats and oils, fish and fishery products, fresh fruits and vegetables, fruit juices, meat and meat products, milk products, and many other products (12). These usually include a detailed description of the product and its composition; limits on contaminants and hazards; weights, measures, and labeling; and methods of sampling and analysis.

The above discussion illustrates the complexity of the information that is made available to authorities charged with controlling food-borne disease. Historically, the standards have been developed to address chemical contamination. Microbial criteria are rarely provided within the Codex system; so far, these criteria are considered to be the domain of national governments and have been based on information about bacterial problems. For instance, guidance on cleaning and disinfection is not specifically developed with viral agents in mind, and the guidelines may be insufficient for application to viruses. Similar observations are applicable to pathogen inactivation steps by heat or other means in processes. It is likely that food-processing technology will be adapted in the near future to better ensure the absence of active viral agents in ready-to-eat foods.

Origins of the Codex Alimentarius
The second half of the 19th century saw the adoption of the first general food laws and the implementation of basic food control systems. During the same period, food chemistry became a reputable discipline. The determination of the "purity" of a food was based primarily on chemical parameters. The concept of "adulteration" was extended to hazardous chemicals in food. Science provided tools to distinguish between safe and unsafe food products.

The different sets of standards arising from the independent development of food laws by different countries gave rise to trade barriers of increasing concern to traders in the early 20th century. Recently, Japan, Singapore, and Hong Kong have started testing incoming shellfish for viruses. This policy could result in unnecessary trade barriers for these products if they originate in countries that apply satisfactory risk management procedures. This situation demonstrates the need to develop internationally accepted criteria for viruses in food matrices.

Trade associations pressured governments to harmonize their various food standards so as to facilitate the trade in safe foods of a defined quality. When FAO and WHO were founded in the late 1940s, there was little, if any, consultation among countries with a view to harmonization.

Growing Role of Science and Increasing Consumer Interest
Meanwhile, rapid progress was being made in food science and technology. With more sensitive analytical tools, people's knowledge about food and associated health hazards also grew, as did interest in food chemistry, food microbiology, and associated disciplines.

In 1953, the WHO World Health Assembly stated that the use of chemicals in food presented a public health problem and that it proposed to conduct relevant studies. As a result, FAO and WHO convened the first joint

FAO/WHO Conference on Food Additives in 1955. This led to the creation of the Joint FAO/WHO Expert Committee on Food Additives (JECFA).

In October 1960, the first FAO Regional Conference for Europe crystallized a widely held view: "the desirability of international agreement on minimum food standards and related questions . . . as an important means of protecting the consumer's health, of ensuring quality and of reducing trade barriers." This eventually led to a resolution, passed during the Eleventh Session of the FAO Conference in 1961, to set up the Codex Alimentarius Commission. In May 1963, the Sixteenth World Health Assembly approved the establishment of the Joint FAO/WHO Food Standards Program and adopted the Statutes of the Codex Alimentarius Commission.

Whereas, previously, consumers' concerns had extended only as far as the "visible problems"—underweight contents, size variations, misleading labeling, and poor quality—they now included a fear of health hazards that could not be seen, smelled, or tasted, such as microorganisms, pesticide residues, environmental contaminants, and food additives. With the establishment of informed consumers' groups, there was growing pressure on governments worldwide to protect communities from poor-quality and hazardous foods.

ROLES OF WHO AND FAO IN RISK ASSESSMENT
Scientific Advice through Expert Bodies
From the very beginning, the Codex Alimentarius was intended to be a science-based activity. The Food Quality and Standards Service of FAO and the Department of Food Safety, Zoonoses, and Food-Borne Diseases of WHO are the lead units responsible for this initiative. JEMRA (the Joint FAO/WHO Expert Meetings on Microbiological Risk Assessment) began in 2000. JEMRA aims to bring together expertise from scientists through the use of the structured microbiological risk assessment (MRA) process as the scientific basis for risk management decisions that address microbiological hazards in foods.

Developing the Risk Analysis Framework
During the past 10 years, the Codex Committee on Food Hygiene (CCFH) has been developing a risk analysis framework, together with FAO, WHO, and individual countries (13). This included the development of new concepts, such as the application of food safety objectives (FSOs), to relate public health goals to the level of stringency required for food safety control. A joint FAO/WHO expert meeting in Kiel, Germany, in 2002 (9) developed guidelines for the application of MRA to the elaboration of standards, guidelines, and related texts. Using these as a basis, FAO/WHO have developed risk

assessments for *Salmonella* in eggs and broiler chickens, *Listeria monocytogenes* in ready-to-eat (RtE) foods, *Vibrio* spp. in seafood, *Campylobacter* spp. in poultry, and other microbial contaminants of food (http://www.who.int/foodsafety/micro/en/). Although the risk assessments have addressed specific questions posed by the Codex, practical guidance for the risk management task of converting the information from these risk assessments into effective risk management strategies still remains to be made final (14).

Why Is the Use of MRA Outputs in Risk Management Complicated?

MRA is a new tool and has only recently been used at the national level, so there is limited experience with its use in different countries. Also, the outputs of MRA are different from those of chemical assessments, which have been under way for many years. The "bright-line" numerical outputs of chemical assessments, often related to a definable minimum dose below which no observable symptoms appear, are used directly to develop standards. With MRA, which deals with living agents reacting differently under different conditions, the assessment is significantly more complex. Compounding this complexity is the fact that for infectious microorganisms no minimum infectious dose applies: even one single organism can sometimes cause disease. Other complicating factors include data gaps in the areas relating to microbial epidemiology and ecology, as well as the complexities of the dose-response relationships, which are influenced by many human and environmental conditions. Finally, regional, cultural, and geographical differences may determine the range of management options that are available and acceptable.

HISTORY OF WORK ON VIRUSES IN THE CODEX

The history of discussions of viruses as food contaminants in the Codex illustrates that this is a process with a long-term timescale. In 1998 at the 31st session of the CCFH, a scientific paper on caliciviruses (16) was presented, with the request that CCFH should consider food safety hazards associated with viruses in order to develop recommendations for their control. The committee recognized that a discussion paper on the subject would be prepared (3). The subsequent *Discussion Paper on Viruses in Foods* (4) provided a comprehensive review of food-borne and waterborne viral gastroenteritis with a focus on caliciviruses and hepatitis, high-risk foodstuffs, virus detection in food and water, and the current status of prevention and disinfection. During the 37th session of the CCFH in Buenos Aires, Argentina, in 2005, an updated discussion paper (6) was presented, and the committee agreed to put viruses on their agenda (5). At its 38th Session, in Houston, Texas, USA,

4 to 9 December 2006, CCFH recognized the need for scientific advice from FAO/WHO. This led to the request to FAO and WHO to jointly organize an Expert Consultation on Foodborne Viruses in 2007 to review the current state of knowledge and to review the availability, feasibility, and practical consequences of using analytical methods for detecting viruses. This meeting, held in May 2007, produced a summary report describing the most important virus-commodity combinations of concern and the different options for management strategies (http://www.who.int/foodsafety/micro/jemra/meetings/virus/en/). This report will be discussed at the next CCFH meeting, which will then decide on future steps. The expert meeting concluded that data at present were insufficient for a full MRA and listed the data gaps to be filled.

HOW CAN RISKS ASSOCIATED WITH VIRUSES IN FOOD BE REDUCED BY APPLYING THE CODEX RISK ANALYSIS FRAMEWORK?

Undertaking an MRA involves using a process to establish the relationship between the frequency and extent of contamination and the incidence and severity of disease. From this relationship, along with knowledge about the dynamics of viral multiplication, survival, and inactivation, the framework calls for the establishment of scientifically defensible metrics (new parameters at relevant points in the food chain) which could subsequently be translated into more traditional measures of food safety control stringency such as process criteria, product criteria, and microbiological criteria. This approach also calls for the clear articulation of practical concepts and methods by which the variability and uncertainty identified in the stochastic assessment of risk can be evaluated and considered in the decision-making process. Therefore, an MRA can be successful only if the scope is well defined (for a specific virus-food commodity), sufficient representative data on the pathogen-commodity combination are available, and—among others—data are available on possible control measures and levels of consumption. The next phase, i.e., to develop effective risk management, requires the selection of proper control measures, including costs and benefits, and data on compliance. It is easy to see that an MRA for viruses will not be available in the near future, but experts in this area should anticipate and prepare for it.

Comparison of MRA of Bacterial and Viral Food-Borne Agents

The work done so far using the MRA framework has focused on bacterial diseases. Some important differences that will have to be considered are that viruses, in contrast to bacteria, will never replicate in food matrices; most food-borne viruses are difficult to detect and quantify due to, among other reasons, the lack of a cell culture system; and dose-response data are very limited.

Very importantly, the efficacy of interventions and control measures (including cleaning and disinfection) is generally not sufficiently well known for viruses in food chains. This points to important areas where data are needed. New analytical tools such as PCR methods to quantify virus particles could speed up the application of MRA to viruses.

Additional Remarks

In the absence of well-defined, representative dose-response information, process modeling can indicate differences in efficacy (in terms of risk reduction), which are defined as relative-risk estimates. If such relative-risk estimates are the only data available, the information can still assist in choosing an adequate intervention scenario, assuming that proper models are constructed and validated. The choice can be between two or more intervention scenarios. It is noted that a scenario often is a combination of more than one intervention. Recent examples of MRAs show that the approach based on relative-risk estimates is valid, e.g., *Enterobacter sakazakii* in infant formula, *Campylobacter* spp. in poultry, and *L. monocytogenes* in RtE food. Probably a first MRA attempt for norovirus (in a defined food matrix) will, at best, aim at such a relative-risk reduction approach, since relevant data are still scarce.

An important element in any future virus MRA/MRM exercise should, where possible, include economic considerations. Costs should be broadly defined and include all types of societal costs.

Bacterial MRA Example: *Listeria monocytogenes* in RtE Food

When considering possibilities for setting up a viral MRA for a certain food category, it may be useful to study the results of the FAO/WHO-organized risk assessment of *L. monocytogenes* in RtE foods (8, 11). Although the risks of virus contamination in food are different in many respects from the risks of bacterial contamination, a parallel is that in many outbreaks of food-borne viral diseases the contamination occurred in the last part of the chain (a clear exception here is the contamination of bivalve mollusks). An interesting aspect of this *L. monocytogenes* MRA is that certain aspects of the study allowed simplification, and consequently, reduction of the models without the necessity to cover the full production chain of the RtE foods.

Food-borne listeriosis is a relatively rare but serious disease with high fatality rates (20 to 30%) compared with other food-borne microbial pathogens. *L. monocytogenes* is an opportunistic pathogen that most often affects those with a severe underlying disease or condition (e.g., immunosuppression, human immunodeficiency virus infection, AIDS, and chronic conditions such as cirrhosis that impair the immune system), pregnant women, unborn or newly delivered infants, and the elderly. *L. monocytogenes* is widely dispersed in the environment and in foods. However, it was not until several large, common-source outbreaks of listeriosis occurred in North America and Europe during the 1980s that the significance of foods as the primary route of transmission of *L. monocytogenes* to humans was recognized (1, 2). An important factor in food-borne listeriosis is that the pathogen can grow to significant numbers at refrigeration temperatures when given sufficient time. Although a wide variety of foods may be contaminated with *L. monocytogenes*, outbreaks and sporadic cases of listeriosis are associated predominantly with RtE foods, a large, heterogeneous category of foodstuffs. The study was limited to foods at retail and their subsequent public health impact at the time of consumption. *Listeria* infection may be invasive or noninvasive. Invasive listeriosis comprises cases when initial infections of the intestinal tissue by *L. monocytogenes* lead to invasion of otherwise sterile body sites, such as the uterus in pregnant women, the central nervous system, the blood, or combinations of these. The incidence rate and factors that govern the onset of the noninvasive form are not known. As a result, this risk assessment considered only invasive listeriosis as the outcome of exposure. The outputs of the MRA were presented in terms of estimates of risk per million servings for the healthy and susceptible populations, which then were used to estimate the number of illnesses in a specified population per year for different food items (Table 1). For milk, for example, the risk per serving was low but the very high rate of consumption resulted in substantial contributions to the total number of predicted cases of illness. In contrast, for smoked fish the risk per serving was estimated to be high but consumption of this product is modest, and, consequently, the total number of cases of listeriosis was moderate. This MRA helped guide management decisions because it showed the food items for which the greatest effect could be expected and showed which possible control measures could be effective. This example il-

Table 1 Mean risk estimates of the number of illnesses per 10 million people per year and the risk per serving for four RtE foods

Food	Cases of listeriosis/10^7 people/yr	Cases of listeriosis/10^6 servings/yr
Pasteurized milk	9.1	0.005
Ice cream	0.012	0.000014
Cold-smoked fish	0.46	0.021
Fermented meat	0.00066	0.0000025

lustrates that, despite major data gaps, an MRA may help focus where control measures could have the biggest impact.

In conclusion, the above overview shows that translating scientific data into defined and measurable control and intervention strategies to reduce the burden of food-borne viral diseases is a time-consuming process. At present, even for the most widely studied food-borne viruses, major data gaps exist. Direct interaction between the research community and risk managers involved in MRA can help focus and prioritize research aimed at advancing the body of knowledge required.

REFERENCES

1. **Bille, L.** 1990. Epidemiology of listeriosis in Europe, with special reference to the Swiss outbreak, p. 25–29. *In* A. J. Miller, J. L. Smith, and G. A. Somkuti (ed.), *Topics in Industrial Microbiology: Foodborne Listeriosis.* Elsevier Science Publishing, New York, NY.

2. **Broome, C. V., B. Gellin, and B. Schwartz.** 1990. Epidemiology of listeriosis in the United States, p. 61–65. *In* A. J. Miller, J. L. Smith, and G. A. Somkuti (ed.), *Topics in Industrial Microbiology: Foodborne Listeriosis.* Elsevier Science Publishing, New York, NY.

3. **Codex ALINORM.** 1999. Codex ALINORM 99/13A, paragraphs 116–118. Food and Agriculture Organization and World Health Organization, Geneva, Switzerland.

4. **Codex Committee on Food Hygiene.** 1999. *Discussion Paper on Viruses in Food* (prepared by The Netherlands with the assistance of Finland, Germany, Italy, and the United States). CX/FH99/1, 32nd CCFH, Washington, DC, 29 Nov to 4 Dec 1999. Food and Agriculture Organization and World Health Organization, Geneva, Switzerland.

5. **Codex Committee on Food Hygiene.** 2005. Report of the 37th session of CCFH. Codex ALINORM 05/28/13, paragraph 168. Food and Agriculture Organization and World Health Organization, Geneva, Switzerland.

6. **Codex Committee on Food Hygiene.** 2005. *Discussion Paper on Virus in Food.* Codex Committee on Food Hygiene, 37th session, Buenos Aires, Argentina, 14 to 19 March 2005. Food and Agriculture Organization and World Health Organization, Geneva, Switzerland.

7. **Codex Committee on Food Hygiene.** 2005. Proposed draft principles and guidelines for the conduct of microbiological risk management (MRM). CX/FH 05/37/6. Codex Committee on Food Hygiene, 37th session, Buenos Aires, Argentina, 14 to 19 March 2005. Food and Agriculture Organization and World Health Organization, Geneva, Switzerland.

8. **Codex Committee on Food Hygiene.** 2005. Proposed draft guidelines on the application of general principles of food hygiene to the control of *Listeria monocytogenes* in ready-to-eat food. CX/FH 05/37/05. Food and Agriculture Organization and World Health Organization, Geneva, Switzerland.

9. **Food and Agriculture Organization and World Health Organization.** 2002. *Principles and Guidelines for Incorporating Microbiological Risk Assessment in the Development of Food Safety Standards, Guidelines and Related Texts.* Joint FAO/WHO Consultation, 18 to 22 March 2002, Kiel, Germany. ISBN 92-5-104845-2. Food and Agriculture Organization and World Health Organization, Geneva, Switzerland.

10. **Food and Agriculture Organization and World Health Organization.** Codex Alimentarius. *Food Hygiene Basic Texts*, 3rd ed., 2003. ISBN 92-5-105106-2. Food and Agriculture Organization and World Health Organization, Geneva, Switzerland.

11. **Food and Agriculture Organization and World Health Organization.** 2004. *Risk Assessment of* Listeria monocytogenes *in Ready-to-Eat Foods.* Microbiological risk assessment series no. 5: Technical report. ISBN 92 4 156262 5. Food and Agriculture Organization and World Health Organization, Geneva, Switzerland.

12. **Food and Agriculture Organization and World Health Organization.** 2006. Format for Codex commodity standards, p. 91. *In Procedural Manual of the Codex Alimentarius Commission*, 16th ed. Food and Agriculture Organization and World Health Organization, Geneva, Switzerland.

13. **Food and Agriculture Organization and World Health Organization.** 2006. Working principles for risk analysis for application in the framework of the Codex Alimentarius, p. 103. *In Procedural Manual of the Codex Alimentarius Commission*, 16th ed. Food and Agriculture Organization and World Health Organization, Geneva, Switzerland.

14. **Food and Agriculture Organization and World Health Organization.** 2006. Report of joint FAO/WHO expert meeting. *The Use of Microbiological Risk Assessment Outputs To Develop Practical Risk Management Strategies: Metrics To Improve Food Safety.* Kiel, Germany, 3 to 7 April 2006. Food and Agriculture Organization and World Health Organization, Geneva, Switzerland.

15. **Food and Agriculture Organization and World Health Organization.** 2006. *Food Safety Risk Analysis—a Guide for National Food Safety Authorities.* Rome 2006. Food and Agriculture Organization and World Health Organization, Geneva, Switzerland.

16. **Koopmans, M.** 1998. *Foodborne Infections by Norwalk-Like Caliciviruses (syn. Small Round Structured Viruses, SRSV).* Codex Food Hygiene Committee Room document 23. Food and Agriculture Organization and World Health Organization, Geneva, Switzerland.

17. **World Health Organization.** 2006. *WHO Consultation To Develop a Strategy To Estimate the Global Burden of Foodborne Diseases: Taking Stock and Charting the Way Forward.* Geneva, 25 to 27 September 2006. WHO/SDE/FOS/2006.1. Department of Food Safety, Zoonoses, and Foodborne Diseases, World Health Organization, Geneva, Switzerland. (http://www.who.int/foodsafety/publications/foodborne_disease/burdensept06/en)

18. **World Health Organization and Food and Agriculture Organization.** 2005. *Understanding the Codex Alimentarius.* Rome 2005. ISBN 92-5-105332-4. World Health Organization and Food and Agriculture Organization, Geneva, Switzerland. (http://www.fao.org/docrep/008/y7867e/y7867e00.htm)

19. **World Trade Organization.** 1994. *SPS Agreement: Agreement on the Application of Sanitary and Phytosanitary Measures.* World Trade Organization, Geneva, Switzerland. (http://www.wto.org/english/docs_e/legal_e/15-sps.pdf)

20. **World Trade Organization.** 1994. *TBT Agreement: Agreement on Technical Barriers to Trade.* World Trade Organization, Geneva, Switzerland. (http://www.wto.org/english/docs_e/legal_e/18-trims.pdf)

Food-Borne Viruses: Progress and Challenges
Edited by Marion P. G. Koopmans, Dean O. Cliver, and Albert Bosch
© 2008 ASM Press, Washington, DC

Risk Assessment of Viruses in Food: Opportunities and Challenges

10

Arie H. Havelaar and Saskia A. Rutjes

Microbiological risk analysis originated in the 1980s, with the publication of a seminal paper on dose-response assessment by Haas (23). Building on this, the first studies concerned the safety of drinking water. Viruses were important target organisms in these first studies. Most early risk assessments focused on enteroviruses and rotaviruses, for which culture methods and dose-response information were available (19, 20, 24, 37). These studies demonstrated that risks of viral contamination can be analyzed by the risk assessment paradigm. Risk assessments in the domain of food safety have focused primarily on bacterial pathogens, for which routine culture methods are generally available; quantitative information on the occurrence of bacteria along the food chain has been produced at an increasing pace. Methods to quantify infectious viruses (and protozoa) in foods are generally more complex or not available at all. This implies that quantitative risk assessment for these organisms is hampered by limited data availability even more than are risk assessments of bacterial pathogens.

This chapter proposes a general framework for risk assessment of viruses in food and evaluates the current data availability. A risk assessment typically consists of four related steps: hazard identification, exposure assessment, hazard characterization, and risk characterization. For a further description of the risk analysis framework, the reader is referred to chapter 9.

Hazard identification in microbial risk assessment is usually defined by the risk management question and is not discussed further. However, expo-

ARIE H. HAVELAAR, Laboratory for Zoonoses and Environmental Microbiology, Centre for Infectious Diseases Control Netherlands, National Institute for Public Health and the Environment, 3720 BA Bilthoven, and Institute for Risk Assessment Sciences, Division of Veterinary Public Health, Faculty of Veterinary Medicine, Utrecht University, PO Box 3508 TD Utrecht, The Netherlands. SASKIA A. RUTJES, Institute for Risk Assessment Sciences, Division of Veterinary Public Health, Faculty of Veterinary Medicine, Utrecht University, PO 3508 TD Utrecht, The Netherlands.

sure assessment, hazard characterization, and risk characterization are discussed below.

GENERAL APPROACH TO EXPOSURE ASSESSMENT

Exposure assessment aims to quantify the exposure of consumers, via a particular food product, to the pathogens of interest. For this purpose, it is necessary to know the probability of occurrence (prevalence) of viruses in a food item at the moment of consumption and the (distribution of the) number of viruses in contaminated units. Additionally, information must be available on the amount of food consumed and the frequency of consumption. Direct measurements of food contamination at the moment of consumption are generally not available; therefore, estimates are usually based on information obtained at earlier stages of the food chain (e.g., at harvest or at retail). For this calculation, a batch of food is defined as being made up of a number of units, e.g., one oyster or one head of lettuce. Units and unit sizes typically change along the food chain; e.g., a lettuce is used to make several servings of a salad. For more details, see reference 39.

A general model for estimation of exposure to protozoa in drinking water was presented by Teunis et al. (54). This model can be adapted for viruses in food:

$$\text{Dose} = P \times C \times 1/R \times I \times 10^{-DR} \times W$$

where Dose is the dose (number of infectious virus particles) ingested in a single meal, P is the prevalence at a specified step in the food chain, C is the measured concentration in contaminated units at that step in the food chain, R is the percent recovery of the detection method, I is the percent infectivity of the detected viruses, DR is the (decimal) reduction of downstream processes, including preparation for consumption, and W is the amount of food consumed (per serving).

QUANTIFYING THE OCCURRENCE OF VIRUSES IN FOOD

The procedure for detection of viruses in food can be divided into three steps: release of virus from the food matrix, concentration of the released virus, and virus detection (see chapter 7). P, C, R and I together reflect the "true" concentration of infectious particles in the food; they depend on the detection procedure applied and are correlated. The prevalence P (i.e., the percentage of units contaminated by one or more infectious particles) acts as a simple multiplier to the other factors in the model. It is the only information that can be obtained by presence-absence tests, but it is not sufficient for a quantitative risk assessment, for which quantitative methods should be applied.

Virus Release

Virus release can be performed either by subjecting the food item to mechanical or enzymatic homogenization or by rinsing it with an appropriate elution buffer, depending on the matrix to be analyzed. Elution of virus and release of PCR inhibitors are dependent on the salt and protein content of the elution buffer and pH, and they determine to a great extent the efficiency of virus detection. Recoveries are determined by the addition of a process control, preferably with the same physiologic characteristics as the target virus and as early as possible in the phase of virus release, to ensure that recoveries of the process control are representative of the target virus. Since the recovery can vary significantly between samples, the use of internal standards with every sample is recommended (33, 48) and is currently discussed by the European Committee for Standardization (CEN) (CEN/TC275—Food analysis—Horizontal methods/WG6—Microbial contamination/TAG4—Detection of viruses in food). This approach, however, does not control the detachment of the virus from the food item and might consequently overestimate R. Detachment is highly dependent on the elution buffer used and characteristics of the virus, such as pI, demonstrating the importance of the choice of a process control virus (11, 22, 56).

Virus Concentration

The need for concentration depends mainly on the volume of buffer used to release the virus from the food matrix. Common concentration procedures include precipitation, ultracentrifugation ($100,000 \times g$), and ultrafiltration, the applicability of all three being dependent on the matrix studied. The remaining concentrates either can be used directly to detect viruses by cell culture methods, or must undergo nucleic acid extraction for detection by reverse transcription-PCR (RT-PCR). Many methods for extraction and reduction of the level of inhibitory substances have been described, with the method based on specific binding of nucleic acids to silica beads, described by Boom et al. (6) as being one of the most commonly used. Recoveries determined by cell culture methods are generally higher than recoveries determined by molecular detection methods, because the latter also detect defective particles in the spike suspension, which might be more easily lost during processing than are intact virus particles (30, 48, 50). This indicates that recoveries determined by molecular methods might be underestimated.

Virus Detection

Viruses can be detected either by culture of virus particles in susceptible cell lines or by molecular detection methods, which are based on amplification of the viral genome. Although different cell culture methods have been described for several enteric viruses (8), no reliable cell culture systems are available for viruses such as hepatitis E virus (HEV) and norovirus (NoV) (14).

Viruses that replicate in monolayer cell lines often cause cytopathic effects (CPE) by virus-specific killing of infected cells. Viruses can be enumerated by estimating the most probable number (MPN) or 50% tissue culture infectious dose (TCID$_{50}$), using liquid cell culture systems or in a plaque assay with a solid medium. By combining cell culture and RT-PCR, the usual long incubation periods for CPE formation are circumvented (10, 13, 40). Moreover, cell culture plus RT-PCR allows the detection of viruses that grow without causing CPE. Cell culture methods detect only virus particles that are able to invade and multiply in host cells; therefore, it can be assumed that $I = 100\%$.

Because of the absence of a robust cell culture system for several viruses, detection of viruses in foods currently focuses on the use of molecular techniques. The most frequently used molecular detection method is RT-PCR, which is based on the specific amplification of conserved regions of the virus genome. The technique can be extremely sensitive and specific; however, amplification can be easily inhibited by substances present in the matrix analyzed. Therefore, the removal or inactivation of potential inhibitors is a major determinant of efficient virus detection (3, 31, 48). Conventional RT-PCR can be used to determine the presence or absence of virus in the analyzed foodstuffs, as well as the numbers of viral genomes present in the undiluted sample by estimating the MPN in PCR-detectable units per gram of analyzed food (35μg). With the introduction of the real-time technologies, the numbers of viral genomes in the original sample could be quantified by comparison with an external standard of serially diluted control DNA and/or RNA. However, if purified nucleic acids, which are more efficiently amplified than nucleic acids isolated from the food samples (which are notorious for the presence of inhibitory substances [33, 48]), are used as standards, the virus concentration in the analyzed food item will be underestimated.

There is no "gold standard" to define 100% detection efficiency. In general, cell culture-PCR provides higher counts than do plaque and TCID$_{50}$ assays and could therefore be considered a de facto standard. A more pragmatic approach would be to consider available dose-response information. If the exposure in the dose-response model is quantified by a cell culture method, it may be appropriate to define this method as the standard.

Molecular methods detect nucleic acids and do not discriminate between viable infectious virus particles and defective noninfectious virus (44); this limits their use for measuring the virological safety of foods. The parameter *I* should therefore represent the fraction of the detected RNA or DNA that originates from infectious particles. *I* depends on the history of the food product before sampling and is expected to decrease with time as a result of inactivation by environmental or food-processing conditions. The ratio between PCR-detectable units and cell culture-detectable units has been reported to increase over time (18) and is highly variable. For enteroviruses in surface waters, the ratio has been shown to vary between 70:1 and 50,000:1 (47). A ratio of 100:1 appears to be a conservative default factor, but the uncertainty in this factor limits the use of molecular data to estimate public health risks and may lead to overestimation of actual risks. Decontamination treatments can markedly affect the ratio of infectious to noninfectious virus particles. Duizer et al. (13) have shown that several treatments (including boiling and UV irradiation) that reduce the concentration of infectious (feline and canine) caliciviruses by at least 5 orders of magnitude are not accompanied by similar decreases in the concentration of PCR-detectable units. For other treatments (including chlorination), the differences were less marked but also significant.

Effect of Virus Aggregation on Different Methods

After being released from the food item, viruses might be present in the extract as aggregates because of hydrophobic interactions between viruses and contaminating substances. These aggregates remain intact under the relatively mild cell culture conditions, indicating that enumeration by cell culture methods such as plaque assays or MPN methods will underestimate the virus concentrations (53). For detection by molecular methods, viral genomes are released in a highly denaturing lysis buffer, which simultaneously separates the aggregated viruses. Before molecular detection, purified nucleic acids can be heated to destroy possible inter- and intramolecular interactions, which subsequently results in single-stranded viral genomes ready for detection. If less than the entire concentrate is analyzed, determination of virus concentrations by both types of detection methods is affected by the presence of virus aggregates in the concentrate; the chance of detection is reduced, implying that virus concentrations will be underestimated. On the other hand, quantitative detection by molecular methods might overestimate the virus concentration in a positive sample.

Proxies

Viruses persist in the environment much longer than bacteria do, suggesting that the presence of fecal indicator bacteria such as *Escherichia coli* is not a suitable basis on which to predict the presence of pathogenic viruses. This is demonstrated by the fact that compliance with bacterial standards for category A shellfish culture areas (less than 230 *E. coli* CFU per 100 g of whole shellfish flesh) does not prevent enteric disease outbreaks associated with the consumption of shellfish (43). Therefore, other indicators for viruses should be considered. The presence of F-specific RNA (F-RNA) bacteriophages has been suggested as an adequate viral indicator on the basis of their size, charge, shape, and stability in the environment (27). Doré et al. (9) have shown that F-RNA bacteriophages were frequently detected at levels exceeding 1,000 PFU per 100 g in oysters that complied with bacterial standards. High levels of F-RNA bacteriophage contamination were strongly associated with fecal pollution of the harvest area and with shellfish-associated disease outbreaks. F-RNA bacteriophage contamination exhibited a marked seasonal trend that was consistent with the trend of oyster-associated gastroenteritis in the United Kingdom. NoV contamination of oysters exhibited a seasonal trend that was consistent with the trend of F-RNA bacteriophage contamination and with the incidence of disease. Myrmel et al. (38) have found a positive correlation between F-RNA phages and NoV in shellfish from Norway, although F-RNA phages were present in only 43% of the NoV-positive samples. Allwood et al. (2) detected F-RNA phages but no NoV RNA in outbreak-associated samples of produce from a food retail establishment.

Alternatively, three commonly occurring human viruses have been evaluated as possible indicators of the presence of human pathogenic viruses: human adenoviruses and polyomaviruses (JCPyV and BKPyV) because of their abundant presence in sewage (1, 42) and human circoviruses, consisting of a group of small DNA viruses that are persistently present in the majority of the population (26, 38). Although results are promising, the suitability of bacteriophages and human viruses as virus proxies in risk assessment models needs further research to evaluate whether a quantitative relationship can be established.

Detection Limit

Sensitivity of detection is one of the most important issues in virus detection in food. Rzezutka et al. (50) were able to detect 10^4 RT-PCR-detectable units of hepatitis A virus (HAV) in 90 g of fresh strawberries after optimization of the method. Thus, absence of an RT-PCR signal indicates that fewer than 3×10^3 RT-PCR-detectable units were present in 25 g of the studied food item; it does not indicate that the virus was absent. Especially when combined foods are studied, recoveries can drop dramatically, leading to higher detection limits (48). The reproducibility of the detection assay increases with increasing virus contamination levels because the chance of PCR amplification increases, as does the chance of investigating a contaminated piece of the food. This has been demonstrated by Dubois et al. (12), who performed an extensive determination of the characteristics of methods for the detection of HAV in inoculated tap waters, salad vegetables, and soft fruits. The detection limit for fruit or vegetables was in the range of 500 to 15,000 RNA copies per 25 g, whereas in water 250 to 2,500 RNA copies per 2 liters were still detectable. Furthermore, molecular detection demands continuous attention because of the high mutation rate associated with the error-prone nature of RNA replication (see chapter 6). The resulting genetic diversity may influence the sensitivities of RT-PCR detection by several log units (29, 57).

OTHER ELEMENTS OF EXPOSURE ASSESSMENT

Reduction by Processing or Storage

Viruses survive well under normal conditions of food storage, and most foods that are associated with viral illness undergo little or no preparation before consumption (e.g., oysters and vegetables). Hence, in many cases, reduction of the concentration of infectious viruses between the step in the food chain where data are collected and the moment of consumption may be negligible. Some foods (e.g., mussels) are lightly cooked before consumption, leading to some reduction in the concentration of infectious viruses. For example, Croci et al. (7) demonstrated that only one of three traditional Italian ways of cooking mussels completely inactivated culturable HAV, whereas the two other methods resulted in considerable (>3 log units) but not complete elimination.

Consumption

The amount (by mass) of products eaten during a meal is usually obtained from food consumption surveys and is similar to that in other microbiological risk assessments. Some foods with high virus risks, such as shellfish, may be consumed infrequently and/or by a limited proportion of the population; hence, it may be more difficult to obtain specific data.

HAZARD CHARACTERIZATION: DOSE-RESPONSE MODELS

A dose-response relationship characterizes the relationship between the ingested number of infectious virus particles and the probability of an adverse health outcome. Usually, a conditional chain of events is considered: exposure → (asymptomatic) infection → illness. Microbial dose-response models are based on three basic assumptions that conceptualize the biological basis of host-pathogen interactions (58): single-hit, independent action, and random distribution. From these assumptions, the single-hit family of models, comprising the exponential model, the hypergeometric model, and the (approximate) Beta-Poisson model, can be derived (23, 52, 59). In the 1950s, several series of experiments were performed with viruses, including several strains of poliovirus, echovirus, and rotavirus (Table 1). The data show that some viruses (in some hosts) are highly infectious (e.g., the probability that a single virus unit can establish infection is greater than 10% for poliovirus strain 1sm in adults, for poliovirus strain 3 Fox in infants and premature neonates, and for rotavirus strain CJN in adults). In contrast, other poliovirus-1 strains are less infectious. These data demonstrate the variability of dose-response relationship, depending on the properties of the virus, the host, and the matrix in which the virus is ingested. This variability makes it difficult to predict the outcome of a single exposure event. However, at the population level, the risk is determined mainly by the occurrence of highly infectious viruses and/or highly susceptible hosts. Hence, use of one of the dose-response relationships with a high single-hit probability is recommended. Several authors

Table 1 Dose-response models for virus infection in human volunteer experiments[a]

Virus species and strain	Exposed population	Model[b]	DR parameters[c]	Low-dose approximation[d]	ID50[e]
Echovirus-12	Healthy adults	BP	$\alpha = 0.401$, $\beta = 227.2$	1.76×10^{-3}	1.05×10^3
Poliovirus					
1 sm	Adults	EX	$r = 0.491$	4.91×10^{-1}	1.411
1 LSc2ab	Infant (newborn)	BP	$\alpha = 0.114$, $\beta = 159.0$	7.2×10^{-4}	6.93×10^4
1	Infant (2 mo)	EX	$r = 9.1 \times 10^{-3}$	9.1×10^{-3}	76.2
3 Fox	Infant	BP	$\alpha = 0.533$, $\beta = 2.064$	0.258	15.513
3 Fox	Infant (premature)	BP	$\alpha = 0.299$, $\beta = 0.552$	0.542	5.05
Rotavirus CJN	Adult males	HG	$\alpha = 0.167$, $\beta = 0.191$	4.66×10^{-1}	11.9

[a]Adapted from reference 59 with permission from Woodhead Publishing Ltd.
[b]EX, exponential model; BP, Beta-Poisson model; HG, hypergeometric model.
[c]The exponential model has one parameter, r; the Beta-Poisson and hypergeometric model have two parameters, α and β.
[d]EX, r; BP, α/β; HG, $\alpha/(\alpha + \beta)$.
[e]EX, $\ln 2/r$; BP, $\beta/(2^{1/\alpha} - 1)$.

have used the dose-response relationship for rotaviruses as a default for human-pathogenic viruses.

There is no published dose-response information for NoV. A proportion of the population is resistant to infection with Norwalk virus, the prototype of NoV genogroup I (GGI), which is associated with the ABO histo-blood group type (28, 34). Subsequent studies (45) have confirmed that individuals with the histo-blood group type B may be protected against infection with NoV GGI. Moreover, recent studies have indicated that secretor-negative individuals, identified by polymorphisms in the *FUT2* gene, are resistant to symptomatic NoV GGII infections (32, 55).

Little attention has been paid to modeling the (conditional) probability of illness, given that infection is a function of the ingested dose. The single-hit models are frequently used to directly model the (unconditional) probability of illness, given a particular dose. In that case, the implicit assumption is that the probability of illness, given the existence of infection, increases with the dose of virus ingested.

RISK CHARACTERIZATION

In the risk characterization step, the information from exposure assessment and dose-response models is combined into a risk estimate. Typically, such estimates are first produced for a single meal and then aggregated over several meals and several individuals to produce an estimate of the average population risk in a particular period (e.g., a year). Most current models assume that subsequent exposures are independent, implying that there is no effect of immunity. If this assumption is not valid, more complex (dynamic) models need to be applied.

Estimates of disease incidence can be extended to estimates of disease burden and costs to provide more information for decision making, e.g., cost-effectiveness or cost-utility analyses. For example, Mangen et al. (unpublished data) estimate that, in The Netherlands, the disease burden of NoV- and rotavirus-associated gastroenteritis was 450 and 370 disability adjusted life years, respectively, whereas the costs of illness were €23 million and €22 million per year, respectively.

PUBLISHED RISK ASSESSMENT STUDIES

An early study of food-borne viruses focused on viral contamination of shellfish (46). Data on the occurrence of culturable enteroviruses in U.S. shellfish were used to estimate an average exposure of 6 PFU per 60-g serving. The dose-response model used was based on rotavirus to characterize a highly infectious virus and echovirus to characterize a moderately infectious virus. The risk of infection per single serving was estimated to range between 1/100 if exposed to a moderately infectious virus and 5/10 if exposed to a highly infectious virus. Such high-risk estimates illustrate the potential health risks associated with consumption of raw shellfish but appear to be higher than actual observed illness risks. A possible explanation would be that consumers are partly protected against illness by immunity due to previous exposures (see below).

Recently published risk assessments of viruses in foods are focused mainly on irrigated crops. These studies did follow the above-described modeling approach to some extent. Since the main focus of these studies was to evaluate criteria for irrigation water, the virus contamination of the food was not directly measured but was modeled as a function of the virus concentration in the water, the attachment of viruses to the crops, and the survival of the viruses during growth and postharvest. Petterson et al. (41) constructed a screening-level risk assessment model for salad crops irrigated with secondary-treated municipal effluent (screening-level risk assessments are used to determine order-of-magnitude risk estimates and to identify the parameters in the model to which the output is most sensitive). Their model was a Monte Carlo simulation, including distributions or point estimates for the concentration of enteroviruses in secondary effluent, the amount of water clinging to lettuce leaves, the decay of viruses on crops and during storage, the quantity of lettuce consumed per meal, and a dose-response relationship (based on rotavirus data). The model did not correct for the recovery (R) of the method to concentrate viruses from effluent. Since tissue culture methods were used to detect the viruses, it was implicitly assumed that the infectivity (I) was 100%. The model was most sensitive to virus decay rates on crops. Hamilton et al. (25) followed a similar approach, evaluating different scenarios for irrigation of broccoli, cabbage, or lettuce. In these studies, many parameter estimates were based on worst-case assumptions, but the authors also failed to correct for the recovery rate of the concentration method. Also, both studies combined data from several authors, obtained from studies in different locations and for different periods, into a generic model and did not attempt to evaluate a specific food production system. Stine et al. (51) experimentally evaluated the transfer of viruses from irrigation water to cantaloupe melons, iceberg lettuce, and bell peppers, as well as the transfer of viruses from the surface of vegetables to fresh-cut surfaces, using coliphage PRD1 as a model organism. These data were combined with new survival data and the annual per capita consumption to estimate the maximum concentration of HAV that would achieve an annual risk of infection of 1 per 10,000 person-years. This risk level was originally proposed by the U.S. Environmental Protection

Agency as a goal for drinking water from surface sources (17) and has subsequently been used to evaluate other risks associated with environmental pathogens.

Risk assessment models for NoV in food have not yet been published. Masago et al. (36) published a risk assessment study of NoV in drinking water in Tokyo, Japan, based on presence/absence testing by RT-PCR. Presence/absence data were transformed into concentration data by assuming a Poisson-lognormal distribution of the viruses in the water. Since no dilution series were investigated, the authors also needed to make assumptions about the standard deviation of the lognormal distribution. An exponential dose-response function was estimated based on a 50% infective dose (ID_{50}) of 10 to100 viable NoV particles. The authors assumed that all detected virus particles were infectious and did not take into account the recovery of the concentration and detection procedures.

SECONDARY TRANSMISSION AND IMMUNITY

Current risk assessment models typically focus on one single exposure event and do not usually take secondary transmission or the effect of previous exposures into account. For many food-borne pathogens, secondary transmission is relatively unimportant. It may lead to some additional cases, but these are typically less common than all index cases; hence, the error involved with disregarding secondary transmission is relatively small. For highly infectious organisms, such as viruses, evaluation of secondary transmission may be more critical, either in specific settings (e.g., health care institutions) or for the general population. Dynamic infectious-disease models have been developed for highly contagious diseases whose main mode of transmission is from human to human (e.g., vaccine-preventable and sexually transmitted diseases). These models typically divide the population into different classes in relation to pathogens, e.g., susceptible, infected (and possibly diseased), and recovered. The models track the evolution of these states over time. Individuals in the recovered state are generally assumed to have developed protective immunity. This introduces a second factor that is generally overlooked in current risk assessment models and may greatly influence actual disease risks. Eisenberg et al. (15, 16) have used dynamic models to analyze the impact of secondary transmission and immunity on environmental transmission of viruses, but the use of such models is not yet common practice.

The protective effect of immunity may be reduced by antigenic variation in virus species that show high mutation rates, such as NoV (see chapter 6). Models to account for antigenic evolution are being developed for influenza virus (4, 5, 21) and may also be applicable for some food-borne viruses.

CONCLUSIONS AND RECOMMENDATIONS

Human health risks posed by viruses in food and drinking water can in principle be modeled by "standard" methods for quantitative microbiological risk assessment (QMRA). However, virus detection methods are highly complex and a large number of factors need to be taken into account to arrive at realistic risk estimates. There are few published studies, and these are based on large numbers of simplifying assumptions. Usually, these assumptions are chosen to arrive at a "worst-case" estimate, i.e., one that is expected to overestimate the public health risk. However, none of the published studies applied correction factors for the recovery of the virus concentration and detection methods, which may result in an underestimation of risks. Due to the high mutation rate of RNA viruses, molecular detection of these viruses demands continuous attention, and methods might need to be updated regularly.

Successful application of the QMRA paradigm depends critically on the availability of robust cell culture methods to analyze the concentration of infectious virus particles in food. Such methods are currently not available for the major food-borne viruses (NoV and HAV). The ratio of infectious to noninfectious virus particles is highly variable, and even though the application of a (conservative) default factor is a theoretical possibility, this is associated with major uncertainty. This problem is not peculiar to quantitative risk assessment but is a general problem in assessing the health risks associated with detection of viral nucleic acid in food samples or the environment.

Better dose-response information would be desirable but is less critical. The assumption of a high infectivity of food-borne viruses (i.e., single-hit probabilities in the order of 10%) may be sufficiently accurate. There is a need to take person-to-person transmission and immunity into account by incorporating dynamic models in the risk characterization stage.

Application of QMRA highlights critical data needs, but the paucity of such data is not restricted to QMRA. Any realistic assessment of the risk of virus contamination of food is subject to the same constraints. QMRA models may help to identify factors to which the risk estimates are most sensitive and thus help to prioritize further experimental and observational studies.

REFERENCES

1. Albinana-Gimenez, N., P. Clemente-Casares, S. Bofill-Mas, A. Hundesz, F. Ribas, and R. Girones. 2006. Distribution of human polyomaviruses, adenoviruses, and hepatitis E virus in the environment and in a drinking-water treatment plant. *Environ. Sci. Technol.* **40:**7416–7422.

2. Allwood, P. B., Y. S. Malik, S. Maherchandani, K. Vought, L. A. Johnson, C. Braymen, C. W. Hedberg, and S. M. Goyal. 2004. Occurrence of *Escherichia coli*, noroviruses, and F-specific coliphages in fresh market-ready produce. *J. Food Prot.* **67:**2387–2390.

3. Atmar, R. L., F. H. Neill, J. L. Romalde, F. LeGuyader, C. M. Woodley, T. G. Metcalf, and M. K. Estes. 1995. Detection of Norwalk virus and hepatitis A virus in shellfish tissues with the PCR. *Appl. Environ. Microbiol.* **61:**3014–3018.

4. Boni, M. F., J. R. Gog, V. Andreasen, and F. B. Christiansen. 2004. Influenza drift and epidemic size: the race between generating and escaping immunity. *Theor. Popul. Biol.* **65:**179–191.

5. Boni, M. F., J. R. Gog, V. Andreasen, and M. W. Feldman. 2006. Epidemic dynamics and antigenic evolution in a single season of influenza A. *Proc. Biol. Sci.* **273:**1307–1316.

6. Boom, R., C. J. Sol, M. M. Salimans, C. L. Jansen, P. M. Wertheim-Van Dillen, and J. Van Der Noordaa. 1990. Rapid and simple method for purification of nucleic acids. *J. Clin. Microbiol.* **28:**495–503.

7. Croci, L., D. De Medici, S. Di Pasquale, and L. Toti. 2005. Resistance of hepatitis A virus in mussels subjected to different domestic cookings. *Int. J. Food Microbiol.* **105:**139–144.

8. De Medici, D., L. Croci, S. Di Pasquale, A. Fiore, and L. Toti. 2001. Detecting the presence of infectious hepatitis A virus in molluscs positive to RT-nested-PCR. *Lett. Appl. Microbiol.* **33:**362–366.

9. Doré, W. J., K. Henshilwood, and D. N. Lees. 2000. Evaluation of F-specific RNA bacteriophage as a candidate human enteric virus indicator for bivalve molluscan shellfish. *Appl. Environ. Microbiol.* **66:**1280–1285.

10. Dubois, E., C. Agier, O. Traore, C. Hennechart, G. Merle, C. Cruciere, and H. Laveran. 2002. Modified concentration method for the detection of enteric viruses on fruits and vegetables by reverse transcriptase-polymerase chain reaction or cell culture. *J. Food Prot.* **65:**1962–1969.

11. Dubois, E., C. Hennechart, N. Deboosère, G. Merle, O. Legeay, C. Burger, M. Le Calvé, B. Lombard, V. Ferré, and O. Traoré. 2005. Intra-laboratory validation of a concentration method adapted for the enumeration of infectious F-specific RNA coliphage, enterovirus, and hepatitis A virus from inoculated leaves of salad vegetables. *Int. J. Food Microbiol.* **108:**164–171.

12. Dubois, E., C. Hennechart, G. Merle, C. Burger, N. Hmila, S. Ruelle, S. Perelle, and V. Ferre. 2007. Detection and quantification by real-time RT-PCR of hepatitis A virus from inoculated tap waters, salad vegetables, and soft fruits: characterization of the method performances. *Int. J. Food Microbiol.* **117:**141–149.

13. Duizer, E., P. Bijkerk, B. Rockx, A. De Groot, F. Twisk, and M. Koopmans. 2004. Inactivation of caliciviruses. *Appl. Environ. Microbiol.* **70:**4538–4543.

14. Duizer, E., K. J. Schwab, F. H. Neill, R. L. Atmar, M. P. G. Koopmans, and M. K. Estes. 2004. Laboratory efforts to cultivate noroviruses. *J. Gen. Virol.* **85:**79–87.

15. Eisenberg, J. N., E. Y. Seto, A. W. Olivieri, and R. C. Spear. 1996. Quantifying water pathogen risk in an epidemiological framework. *Risk Anal.* **16:**549–563.

16. Eisenberg, J. N., J. A. Soller, J. Scott, D. M. Eisenberg, and J. M. Colford, Jr. 2004. A dynamic model to assess microbial health risks associated with beneficial uses of biosolids. *Risk Anal.* **24:**221–236.

17. **Environmental Protection Agency.** 1989. National primary drinking water regulations: filtration, disinfection; turbidity, *Giardia lamblia*, viruses, *Legionella* and heterotrophic bacteria. Final rule. *Fed. Regist.* **54:**27486.

18. **Gassilloud, B., and C. Gantzer.** 2005. Adhesion-aggregation and inactivation of *Poliovirus 1* in groundwater stored in a hydrophobic container. *Appl. Environ. Microbiol.* **71:**912–920.

19. **Gerba, C. P., J. B. Rose, C. N. Haas, and K. D. Crabtree.** 1996. Waterborne rotavirus: a risk-assessment. *Water Res.* **30:**292–92940.

20. **Gerba, C. P., and C. N. Haas.** 1988. Assessment of risks associated with enteric viruses in contaminated drinking water, p. 489–494. *In* J. J. Lichtenberg, J. A. Winter, C. I. Weber, and L. Fradkin (ed.), *Chemical and Biological Characterization of Sludges, Sediments, Dredge Spoils and Drilling Muds.* American Society for Testing and Materials, Philadelphia, PA.

21. **Gog, J. R., and J. Swinton.** 2002. A status-based approach to multiple strain dynamics. *J. Math. Biol.* **44:**169–184.

22. **Gulati, B. R., P. B. Allwood, C. W. Hedberg, and S. M. Goyal.** 2001. Efficacy of commonly used disinfectants for the inactivation of calicivirus on strawberry, lettuce, and a food-contact surface. *J. Food Prot.* **64:**1430–1434.

23. **Haas, C. N.** 1983. Estimation of risk due to low doses of microorganisms: a comparison of alternative methodologies. *Am. J. Epidemiol.* **118:**573–582.

24. **Haas, C. N., J. B. Rose, C. Gerba, and S. Regli.** 1993. Risk assessment of virus in drinking water. *Risk Anal.* **13:**545–552.

25. **Hamilton, A. J., F. Stagnitti, R. Premier, A. M. Boland, and G. Hale.** 2006. Quantitative microbial risk assessment models for consumption of raw vegetables irrigated with reclaimed water. *Appl. Environ. Microbiol.* **72:**3284–3290.

26. **Haramoto, E., H. Katayama, O. Kumiko, H. Yamashita, E. Nakajima, and S. Ohgaki.** 2005. One-year monthly monitoring of Torque teno virus (TTV) in wastewater treatment plants in Japan. *Water Res.* **39:**2008–2013.

27. **Havelaar, A. H., M. van Olphen, and Y. C. Drost.** 1993. F-specific RNA bacteriophages are adequate model organisms for enteric viruses in fresh water. *Appl. Environ. Microbiol.* **59:**2956–2962.

28. **Hutson, A. M., R. L. Atmar, D. Y. Graham, and M. K. Estes.** 2002. Norwalk virus infection and disease is associated with ABO histo-blood group type. *J. Infect. Dis.* **185:**1335–1337.

29. **Jothikumar, N., J. A. Lowther, K. Henshilwood, D. N. Lees, V. R. Hill, and J. Vinjé.** 2005. Rapid and sensitive detection of noroviruses by using TaqMan-based one-step reverse transcription-PCR assays and application to naturally contaminated shellfish samples. *Appl. Environ. Microbiol.* **71:**1870–1875.

30. **Kostenbader, K. D., Jr., and D. O. Cliver.** 1973. Filtration methods for recovering enteroviruses from foods. *Appl. Microbiol.* **26:**149–154.

31. **Kreader, C. A.** 1996. Relief of amplification inhibition in PCR with bovine serum albumin or T4 gene 32 protein. *Appl. Environ. Microbiol.* **62:**1102–1106.

32. **Larsson, M. M., G. E. Rydell, A. Grahn, J. Rodriguez-Diaz, B. Akerlind, A. M. Hutson, M. K. Estes, G. Larson, and L. Svensson.** 2006. Antibody prevalence and titer to norovirus (genogroup II) correlate with secretor (FUT2) but not with ABO phenotype or Lewis (FUT3) genotype. *J. Infect. Dis.* **194:**1422–1427.

33. Le Guyader F., A. C. Schultz, L. Haugarreau, L. Croci, L. Maunula, E. Duizer, F. Lodder-Verschoor, C. H. von Bonsdorff, E. Suffredini, W. H. M. van der Poel, R. Reymundo, and M. Koopmans. 2004. Round-robin comparison of methods for the detection of human enteric viruses in lettuce. *J. Food Prot.* **67:**2315–2319.

34. Lindesmith, L., C. Moe, S. Marionneau, N. Ruvoen, X. Jiang, L. Lindblad, P. Stewart, J. LePendu, and R. Baric. 2003. Human susceptibility and resistance to Norwalk virus infection. *Nat. Med.* **9:**548–553.

35. Lodder, W. J., and A. M. de Roda Husman. 2005. Presence of noroviruses and other enteric viruses in sewage and surface waters in The Netherlands. *Appl. Environ. Microbiol.* **71:**1453–1461.

36. Masago, Y., H. Katayama, T. Watanabe, E. Haramoto, A. Hashimoto, T. Omura, T. Hirata, and S. Ohgaki. 2006. Quantitative risk assessment of noroviruses in drinking water based on qualitative data in Japan. *Environ. Sci. Technol.* **40:**7428–7433.

37. Mena, K. D., C. P. Gerba, C. N. Haas, and J. B. Rose. 2003. Risk assessment of waterborne coxsackievirus. *J. Am. Water Works Assoc.* **95:**122–133.

38. Myrmel, M., E. M. Berg, E. Rimstad, and B. Grinde. 2004. Detection of enteric viruses in shellfish from the Norwegian coast. *Appl. Environ. Microbiol.* **70:**2678–2684.

39. Nauta, M. J. 2007. The modular process risk model (MPRM): a structured approach to food chain exposure assessment, p. 99–136. *In* D. W. Schaffner (ed.), *Microbial Risk Analysis of Foods.* ASM Press, Washington, DC.

40. O'Mahony, J., M. O'Donoghue, J. G. Morgan, and C. Hill. 2000. Rotavirus survivial and stability in foods as determined by an optimised plaque assay procedure. *Int. J. Food Microbiol.* **61:**177–185.

41. Petterson, S. R., N. J. Ashbolt, and A. Sharma. 2001. Microbial risks from wastewater irrigation of salad crops: a screening-level risk assessment. *Water Environ. Res.* **73:**667–672.

42. Pina, S., M. Puig, F. Lucena, J. Jofre, and R. Girones. 1998. Viral pollution in the environment and in shellfish: human adenovirus detection by PCR as an index of human viruses. *Appl. Environ. Microbiol.* **64:**3376–3382.

43. Potasman I., A. Paz, and M. Odeh. 2002. Infectious outbreaks associated with bivalve shellfish consumption: a worldwide perspective. *Clin. Infect. Dis.* **35:**921–928.

44. Richards, G. P. 1999. Limitations of molecular biological techniques for assessing the virological safety of foods. *J. Food Prot.* **62:**691–697.

45. Rockx, B. H., H. Vennema, C. J. Hoebe, E. Duizer, and M. P. Koopmans. 2005. Association of histo-blood group antigens and susceptibility to norovirus infections. *J. Infect. Dis.* **191:**749–754.

46. Rose, J. B., and M. D. Sobsey. 1993. Quantitative risk assessment for viral contamination of shellfish and coastal waters. *J. Food Prot.* **56:**1043–1050.

47. Rutjes, S. A., R. Italiaander, H. H. J. L. van den Berg, W. J. Lodder, and A. M. de Roda Husman. 2005. Isolation and detection of enterovirus RNA from large-volume water samples by using the NucliSens miniMAG system and real-time nucleic acid sequence-based amplification. *Appl. Environ. Microbiol.* **71:**3734–3740.

48. Rutjes, S. A., F. Lodder-Verschoor, W. H. M. Van der Poel, Y. T. H. P. Van Duijnhoven, and A. M. De Roda Husman. 2006. Detection of noroviruses in foods: a study on virus extraction procedures in foods implicated in outbreaks of human gastroenteritis. *J. Food Prot.* **69:**1949–1956.

49. Rutjes, S. A., H. H. J. L. van den Berg, W. J. Lodder, and A. M. de Roda Husman. 2006. Real-time detection of noroviruses in surface water by a broadly reactive nucleic acid based amplification assay. *Appl. Environ. Microbiol.* **72:**5349–5358.

50. Rzezutka, A., M. Alotaibi, M. D'Agostino, and N. Cook. 2005. A centrifugation-based method for extraction of norovirus from raspberries. *J. Food Prot.* **68:**1923–1925.

51. Stine, S. W., I. Song, C. Y. Choi, and C. P. Gerba. 2005. Application of microbial risk assessment to the development of standards for enteric pathogens in water used to irrigate fresh produce. *J. Food Prot.* **68:**913–918.

52. Teunis, P. F., and, A. H. Havelaar. 2000. The beta Poisson dose-response model is not a single-hit model. *Risk Anal.* **20:**513–520.

53. Teunis, P. F. M., W. J. Lodder, S. H. Heisterkamp, and A. M. de Roda Husman. 2005. Mixed plaques: statistical evidence how plaque assays may underestimate virus concentrations. *Water Res.* **39:**4240–4250.

54. Teunis, P. F. M., G. J. Medema, L. Kruidenier, and A. H. Havelaar. 1997. Assessment of the risk of infection by *Cryptosporidium* or *Giardia* in drinking water from a surface water source. *Water Res.* **31:**1333–1346.

55. Thorven, M., A. Grahn, K. O. Hedlund, H. Johansson, C. Wahlfrid, G. Larson, and L. Svensson. 2005. A homozygous nonsense mutation (428G→A) in the human secretor (FUT2) gene provides resistance to symptomatic norovirus (GGII) infections. *J. Virol.* **79:**15351–15355.

56. Vega, E., J. Smith, J. Garland, A. Matos, and S. D. Pillai. 2005. Variability of virus attachment patterns to butterhead lettuce. *J. Food Prot.* **68:**2112–2117.

57. Vennema, H., E. de Bruin, and M. Koopmans. 2002. Rational optimization of generic primers used for Norwalk-like virus detection by reverse transcriptase polymerase chain reaction. *J. Clin. Virol.* **25:**233–235.

58. World Health Organization and Food and Agriculture Organization. 2003. *Hazard Characterization of Pathogens in Food and Water—Guidelines.* Geneva, Rome: World Health Organization, Food and Agricultural Organization of the United Nations, Geneva, Switzerland.

59. Zwietering, M. H., and A. H. Havelaar. 2006. Dose-response relationships and foodborne disease, p. 422–439. *In* M. Potter (ed.), *Food Consumption and Disease Risk.* Woodhead Publishing, Cambridge, United Kingdom.

Index

A

Adenoviridae, 126–127, 135
Adenoviruses, human, 226
Alcohol hand gels, 48
Anellovirus, 127
Animal reservoirs of infectious disease, 122
Anti-HAV antibodies, detection, 172–173
Astroviridae, 126–127, 135
Astroviruses, 30
Avian HEV, 74–75
Avian influenza, 133–134
Avian influenza virus, 117, 126

B

Basic reproductive ratio, 158
Becovirus, 31
Biphasic milk fever, 132
Birnaviridae, 127, 135
Blood serum, for culture of cells, 4
Bocavirus, 128–129
Bottleneck events, 156–157
Bovine viral diarrhea, 132
"Brute-force" ultracentrifugation, 10
Bovine spongiform encephalopathy ("mad cow disease"), 2, 13, 14

C

Caliciviridae, 31, 48, 126–127, 135, 149, 179, 180
Caliciviruses, 171
  human, 30–32
Canine calicivirus, 48

Cat-Floc, 9–10
Cell culture, in detection of viruses, 224
  media for, 3–4
  vessels for, 3
Cell culture propagation, of hepatitis A virus, 171–172
Cells, kidney as source of, 4
Centrifugation, 8–9
Chromatography, column, 7
  "membrane," 7
Circoviridae, 127, 135
Circovirus, 127
Circoviruses, human, 226
Creutzfeldt-Jakob disease, 2, 14
Cliver, Dean O., 16
Codex Alimentarius Commission, 214
Codex Alimentarius ("food code"), 210–214
  and international food trade, 210–212
  origins of, 213
Codex Committee on Food Hygiene (CCFH), 210, 214–215
Codex General Principles of Food Hygiene, 212–213
Codex risk analysis, in reduction of viruses in food, 216–219
  risks of viruses in food and, 209–220
  work on viruses in, history of, 215–216
Codex system, 210–212
  commodity standards and, 212–213
  guidelines of, 212
  standards and codes of practice of, 212–213
Contagium vivum fluidum, 1

Conventional reverse transcription-PCR (RT-PCR) assays, 32
Cooking, to inactivate viruses, 199–201
Coronaviridae, 130–131
Coronaviruses, fecal-oral transmission of, 157–158
Crassostrea gigas, 194
Crassostrea virginica, 194
Cytopathic effect, 5

D
Dack, Gail M., 2
Delphi technique, 104
Dengue virus, 132
Depuration, 71, 193, 194
Derjaguin-Landau-Verwey-Overbeek theory of colloidal stability, 191–192
Diarrhea, bovine viral, 132
Diarrheal disease, investigation of, 107
norovirus in stools in, 43
Disease burden, population-based approaches for determining, 97–102
prospective studies of, 98–101
retrospective surveys of, 98, 99
underreported, estimation of burden of, 87–115
Disinfectants, against NoVs, 48
DNA, viral, 1
mutation of, 151–153
DNA virus replication, inhibition of, 7
DNA viruses, 156
Dose-response models, 228–229
Dose-response relationship, for viral infections, 228–229
Drinking water, contamination of, 176
protozoa in, 222

E
Ecological factors, viral disease and, 122
Economic assessments, of burden of AGE and/or NoV, 101–102
Electron microscopy, 1
to detect noroviruses, 89
Encephalitis, tick-borne, 17
Encephalomyocarditis virus (EMCV), 179
Encephalopathies, spongiform, 14
Enteric viral infections, 29
Enteric viruses, 7
Environmental factors, effects on inactivation of virus, 196–197

Enzyme immunoassays, 33–34
Epidemiologic studies, analytical, of food-borne viral gastroenteritis, 102–104
European Agency for the Evaluation of Medicinal Products, 179
European Committee for Standardization of Horizontal Methods for the Molecular Detection of Viruses in Food, 182
Expert opinion, for analysis of food-borne viral gastroenteritis, 104–105
Exposure assessment, in risk assessment, 222
virus survival in food storage, 227

F
Fecal-oral transmission, 17, 18
Feline calicivirus (FCV), 48, 180
Filtration, 8–9
Flaviviridae, 132–133
Flavivirus, 132
Food(s), contaminated, norovirus burden in, 105–106
traceback of, 122
contamination of, at source, 159
by food handler, 159
environmental sources of, 189, 190
effects of compounds on viral growth, 196
exports from different countries, 121
gamma irradiation of, 199
global trade of, 121–122
HAV entry and concentration in, 71
preharvest and postharvest contamination of, 176
processing and storage of, virus survival and, 227
prospective versus retrospective analysis of, 176–178
ready-to-eat, contamination of, 190, 271–218
real-time RT-PCR-based analysis of, 179
transmission of HEV via, 78–79
transmission of viruses by, factors influencing, 189
treatment against NoVs, 49
treatment to remove or inactivate viruses, 198–201
viral genome copies in, estimation of, 181
virus binding to, 190–195
probability of, 191

virus detection in, 171–188
 molecular approaches to, 173–176, 182
 procedures for, 222–227
 standardization of, 178
virus persistence in, consequences of,
 201–202
virus transmission via, 156–157
viruses in, cooking to inactivate, 199–201
 hazard characterization of, 228–229
 risk assessment of, 221–236
viruses on and in, binding and inactivation of, 189–208
zoonotic virus transmission via, 157–158
Food additives, FAO/WHO Conference on
 (1955), 214
Food and Agriculture Organization (FAO),
 food-borne illnesses and, 209
 role in risk assessment, 214–215
Food and Agriculture Organization
 (FAO)/World Health Organization
 (WHO), Conference on Food
 Additives, 213–214
 Expert Committee on Food Additives
 (JECFA), 214
 Expert Consultation on Food-Borne
 Viruses, 216
Food-borne disease, causes of, 6–7
 economic burden of, 46–47
 historic summary of, 19
 NoVs as cause of, 44
 difficulty in diagnosis of, 44–46
Food-borne viral disease, burden of, improving assessment of, 106–108
 detection of, surveillance systems for, 135
 emerging, 117–145
 food handler transmission of, 124
 laboratory and, 89–90
 laboratory-based surveillance for, 93–96
 pathogen-related factors in, 122–123
 person-to-person transmission of, 104
 physician and health care system and, 88–
 89
 public health system and, 90–91
 surveillance of, challenges in, 87–92
 surveillance pyramid and, 91, 92
 transmission of, control of, 123
 prerequisites for, 123–124
 virus and patient and, 87–88
 zoonotic transmission of, 123–124
 factors favoring, 126
Food-borne viruses, Codex risk analysis and,
 209–220
 detection of, 8–12, 80, 171–188

effect of virus aggregation in, 225
 proxies and, 226
 sensitivity of detection and,
 226–227
 steps in, 222
 virus aggregation and, 225
 virus concentration in, 223–224
 virus detection in, 224–225
 virus release in, 223
epidemiology of, viral evolution and
 relevance for, 147–169
 hazard characterization of, 228–229
 history of cases of, 2
 identification of, 8–12
 irrigated crops and, 230
 obtaining samples for testing, 8
 risk assessment of, 221–236
 seeding event, 158
 state of art, 29–64
 transmission of, prevention of, 80
Food chain surveillance, integrated, 97
Food chemistry and microbiology, consumer
 interest in, 213–214
Food Research Institute of University of
 Chicago, 2
Food trade, and Codex Alimentarius,
 210–212
Food virology, historic overview of, 1–28
Foot-and-mouth disease, 126
Freon extraction method, 9

G

Gamma irradiation of food, 199
Gastroenteritis, acute, economic burden of,
 101–102
 population-based prospective studies
 of, 98–101
 population surveys of, 106–107
 retrospective surveys of, 98, 99
 surveillance for, 92–97
 food-borne outbreaks of, due to
 noroviruses, 96
 viral, 29, 30–31
 food-borne, analytical epidemiologic
 studies of, 102–104
 assessment of proportion of,
 102–105
 expert opinion for analysis of,
 104–105
 public health system and, 90–91
 norovirus infection causing, 44–46
 noroviruses in, 43–44

Gastrointestinal illness, food-borne viruses causing, 30
Genes, encoding proteins of viruses, 149–150
Genome copies, infectivity and, 182
Green onions, hepatitis A virus and, 193
Gyrovirus, 127

H

HAV genotyping, to detect HAV, 73
Hazard Analysis and Critical Control Point (HACCP) system, 177, 212
application to viral safety, steps in, 177
Health care system, food-borne viral disease and, 88–89
Hedra viruses, 134
HeLa cell line, 4
Henipavirus, 134
Hepacivirus, 132
Hepatitis, acute sporadic, HEV infection and, 76
enterically transmitted, 65–85
food-borne outbreaks of, 2
food-borne viruses causing, 30
infectious, history of transmission of, 6
viral, clinical presentation of, 65
viruses causing, 65, 66
Hepatitis A infection, 6
age at and clinical expression of, 68
antibodies protective against, 69–70
diagnosis of, 69
epidemiological patterns of, 67–69
green onion consumption and, 193
immunization against, 70
routes of transmission of, 69
Hepatitis A virus, 17–19, 124, 149, 156, 171
contamination of foods, detection, 72–73
cooking to eliminate, 199–201
effect of temperature on, 195–196
entry and concentration in foods, 71
epidemiology of, 67–69
food-borne outbreaks due to, 172
food-borne transmission of, 70–71
prevention of, 71–72
genotypes of, 67
modified atmospheric packaging and, 198
multiplication and excretion of, 67
thermal and environmental stability of, 70
transmission of, 38, 135

vaccination for postexposure prophylaxis, 73
virology and molecular biology of, 65–67
Hepatitis E infection, clinical features and outcome of, 76–77
diagnosis of, 77
endemic, differences in areas with and without, 79–80
epidemiology of, 75–76, 79
person-to-person transmission of, 75–76
protective antibodies against, 77
transmission of, evidence of, 78–79
Hepatitis E virus, detection in food, 80
genotypes of, 74–75
in meat, 126
outbreaks and spread of, 120–121
thermal and environmental stability of, 77
virology and molecular biology of, 73–74
Hepatitis viruses, enterically transmitted, 80–81
Hepatomegaly-splenomegaly syndrome, HEV and, 75
Hepatovirus, 149
Hepeviridae, 126–127, 135
High-hydrostatic-pressure processing (HHP), for virus inactivation, 198–199
Histo-blood group antigens, 36–37
in shellfish, 194
Hog cholera virus, 126
Human immunodeficiency virus, 117–118
Hydrostatic pressure, to inactivate HAV, 72

I

Immune electron microscopy, to detect HAV, 72
Immunity, protective effect of, risk assessment models and, 231
Immunoassays, enzyme, 33–34
Infections, ecological factors and, 122
viral. *See* Viral infections
Infectious disease models, 231
Influenza A viruses, 133–134
Influenza virus, transmission of, 135
Irrigated crops, and viruses in foods, 230

J

Japanese encephalitis virus, 132

K

Kapikian, Albert, 30–31
Kidney, as source of cells, 4
Kuru, 13

L

Laboratory, examination of stool samples, 89
food-borne viral disease and, 89–90
Laboratory-based surveillance, 93–96
Lagovirus, 31, 149
Leviviridae, 179
Listeria monocytogenes, 217–218
Listeriosis, food-borne, 218
"Live-virus" vaccines, 4
Liver disease, acute-on-chronic, 77

M

Mad cow disease, 13, 14
Mammalian viruses, taxonomic groupings of, 124, 125
Marination, 199
Meat, hepatitis E virus in, 126
Medium 199, 3
"Membrane chromatography," 7
Microbial dose-response models, 228–229
Microbiological risk assessment, bacterial example of, 217–219
    JECFA and, 214
    of bacterial and viral food-borne agents, compared, 216–217
    outputs in risk management, 215
Microbiological risk management, 210, 211
Microscope, for demonstration of virus replication, 5
Migration, introduction of disease due to, 120
Mimiviruses, 148
Modified atmospheric packaging, 198
Molecular biology, 29
Molecular techniques, in detection of viruses in foods, 182, 224–225
Molluscan shellfish, 123
Mouse encephalomyelitis virus, 192
Murine NoV-1, 179
Mutagenesis, lethal, mechanism of, 162–163
Mutation, and role in virus evolution, 151–153
    genetic variation of viruses and, 150

N

Nabovirus, 31
National health help line, for syndromic surveillance, 92–93
Nipah virus, 117, 122, 124, 134–135
Norovirus, 31, 124, 149
    genogroups and genotypes of, 194
Norovirus infection, as outbreak-related disease, 96
    causing large outbreaks of gastroenteritis, 44–46
    economic burden of, 101–102
    endemic, epidemiology of, 43
    food-borne outbreaks of, foods implicated in, 38–39
    from NoV-contaminated environments, 39, 40
    illnesses attributed to, 34–35
    immunity to, 36–38
    molecular epidemiology of, 41
    pathogenesis of, 35–37
    population-based prospective studies of, 98–101
    presentation of, 34–35
    prevention and control of, 47–49
    questions about mechanisms of, 37
    saliva samples to estimate, 107
    "sporadic," prevalence estimates for, 93, 94–95
    sporadic disease due to, 41–44
    symptoms of, 34
    transmission modes and settings of, 44–46, 135
    treatment of, 47
    waterborne outbreaks of, 39
Noroviruses, 2, 7, 30, 149, 171
    binding patterns of strains of, 37
    cell culture and animal models of, 33–34
    classification of, 31–32
    control of, 47–49
    detection in clinical samples, 32–34
    diagnostic laboratories and, 89
    disinfection against, 47–49
    diversity of, 175–176
    enzyme immunoassays for, 33–34
    food-borne, estimation of burden of, 103
        outbreak data to estimate, 103, 105
        outbreaks due to, 172
        outbreaks of gastroenteritis due to, 96
    gaps in knowledge concerning, 49–51
    genotype IIc, 158
    in contaminated foods, 105–106
    in gastroenteritis, role of, 43–44

Noroviruses, *(continued)*
  recombination in evolution of, 154
  risk assessment models for, 230–231
  RT-PCR assays for, 32–33, 89–90
  secondary person-to-person transmission
    of, 104
  subclades and variants of, 41, 42
  transmission of, 38–41
    person to person, 96
  treatment of food against, 49
  vaccine against, obstacles to development
    of, 49
Norwalk virus, 13, 31, 36–37
NoV genome, detection of, 35–36
Nucleic acid sequencing, to detect HAV, 73
Nucleic acid type viruses, 7–8

O

Open reading frames (ORFs), 31
"Organic flocculation," 11
Orthomyxoviridae, 133–134
Outbreak data, to estimate proportion of
    food-borne NoV, 103, 105
Outbreak surveillance, 96
Oysters, virus elimination by, 194
  virus persistence in, 198

P

Papillomaviridae, 128, 135
Paramyxoviridae, 134–135
Parvoviridae, 128–129, 135
PCR, quantitative, to assess viral load of
    NoV, 44
Pestivirus, 132
Pharmacy, for syndromic surveillance, 92–93
Physician, food-borne viral disease and,
    88–89
Picornaviridae, 126–127, 135, 149, 179
Picornaviruses, mutation and rates of evolu-
    tion of, 152
  translation of, 173–176
Plants, virus contamination of, 192–193
Poison control center database, for syn-
    dromic surveillance, 93
Poliomyelitis, food-borne, 2
  recombinants of poliovirus vaccine and
    enteroviruses and, 154
Poliovirus 1, inactivation of, 12
Poliovirus type 1 strain, 192
Poliovirus vaccines, 17

Polyethylene glycol (PEG), dialysis against,
    11
Polyethylene glycol (PEG) 6000, 11
Polymer two-phase system, 11
Polyomaviridae, 129–130, 135
Polyomaviruses, 226
Population, as basis for determining disease
    burden, 97–102
Population increases, risk of infectious dis-
    ease in, 120–121
Population surveys, of acute gastroenteritis,
    106–107
Primary monkey kidney (PMK) cell
    cultures, 11–12
Prion(s), 2
  description of, 13
Prion diseases, 13–15
Produce, fresh, contamination of, 176
  NoV-contaminated, 106
Proofreading-repair activity, 151
Prospective cohort studies, population-
    based, of AGE or NoV infection,
    98–101
Proteins, of viruses, genes encoding,
    149–150
Protozoa, in drinking water, model for
    estimation of, 222
PrP^c, 13
PrP^{Sc} (infective prion), 13, 14–15
Public health system, food-borne viral dis-
    ease and, 90–91

Q

Quality assurance and quality control
    (QA/QC) measures, 178, 180
Quasispecies, viral, biological implications
    of, 161–163
  concept of, 159–161

R

Real-time reverse transcription-PCR
    assays, 32, 33, 179
Recombination, and virus evolution,
    153–154
  definition of, 153
Relaying, 193
Reoviridae, 126–127, 135
Respiratory tract viral pathogens, 163–164
Retrospective studies, of acute gastroenteri-
    tis, 98, 99

Reverse transcription-PCR assays, 32–33, 43, 46
  in detection of virus concentration, 223, 224
  to detect HAV, 72
    in food, 173–176
  to detect noroviruses, 89–90
    in food, 173–176
Rhinoviruses, 8
Risk analysis, Codex. *See* Codex risk analysis
  components of, 209
  developing framework of, 214–215
Risk assessment, early, of enteroviruses and rotaviruses, 221
  exposure assessment in, 222
  four steps in, 221
  in risk analysis, 210
  of viruses in food, 221–236
    bacterial pathogens and, 221
  WHO and FAO in, 214–215
Risk assessment studies, published, 229–231
  secondary transmission and immunity in, 231
Risk characterization, 229
Risk estimate, 229
Risk management, microbiological, 210, 211, 214, 215
RNA-containing capsids, noninfectious unaltered, 182
RNA viruses, 1, 122, 156, 173
  intact, 8
  mutation of, 151–153
  variability of, 164
Rotaviruses, 30, 122–123

S

Sabin, A.B., 4
Salad crops, irrigated, concentration of viruses on, 230
Saliva samples, to estimate incidence of NoV infection, 107
Sanitizers, 48
Sapovirus, 31, 48, 149
Sapoviruses, 30
  classification of, 32
SARS coronavirus (CoV), 126, 130–131
  food-related transmission of, 131
  transmission of, 135
Scrapie, 13
Seafood, NoV-contaminated, 106

Serologic studies, limitations of, 43
Severe acute respiratory syndrome, 121
Severe acute respiratory syndrome virus, 117
  origin and transmission of, 157–158
Shellfish, contamination of, 176, 190
  cooking to inactivate viruses, 199–201
  culturable enteroviruses in, 229–230
  HBGA A-like carbohydrate in, 194
  marination of, 199
  persistence of viruses in, 197–198
  uptake and concentration of viruses in, 194
  virus inactivation in, physical factors influencing, 200–201
  virus uptake by, 193–195
Southampton virus, 31
Spongiform encephalopathies, 14
Stool sample, laboratory examination of, 89
  submission to physician, 88
Surveillance, active, 97
  integrated food chain, 97
  laboratory-based, 93–96
  outbreak, 96
  syndromic, motivation and compliance in, 93
    national health help line for, 92–93
    pharmacy-based, 92–93
    poison control center database-based, 93
    sources of data for, 92
Surveillance data, interpretation of, 97
  sources of, making better use of, 108
Surveillance pyramid, food-borne viral disease and, 91, 92
Swine fever virus, 132
Swine HEV, 74–75

T

Telephone triage systems, nurse-led, 88–89
Temperature, effects on inactivation of virus, 195–196
Tick-borne encephalitis, 17
Tick-borne encephalitis virus, 132
Transmission electron microscopy, 8

U

Ultracentrifuge, 10–11
U.S. Environmental Protection Agency, 5

U.S. Food and Drug Administration, 179
UV irradiation of shellfish, 72

V
Vaccine(s), against NoVs, obstacles to development of, 49
    from kidneys of monkeys, 4
    HAV, 70
    "live-virus," 4
    oral vaccine, 4
    poliovirus, 17
    recombinant HEV, 80
vCJD, 14–15
Vesivirus, 31, 48, 149
Viral disease, emergence and spread of, factors contributing to, 119–123
    food-borne. *See* Food-borne viral disease
Viral genome copies in food, estimation of, 181
Viral infections, dose-response relationship for, 228–229
    emergence and spread of, factors contributing to, 118–119
        human factors in, 120–121
    emerging, 117–119
    enteric, 29
Viral quasispecies, biological implications of, 161–163
    concept of, 159–163
Virogenomics, 149
Virology, food, historic overview of, 1–28
Virus(es), adaptation to changing environments, 147
    aggregation of, effect on virus detection, 225
    binding to foods, 190–195
        probability of, 191
    causing hepatitis, characteristics of, 66
    characterization of, 6
    classification of, 7
    Codex risk analysis of, history of, 215–216
    contamination of plants by, 192–193
    detection of, 3
    DNA or RNA in, 1
    enteric, binding to foods, 202
    evolution of, 148–150
        in vivo, dynamics of, 155–163
        mutation and, 151–153
        recombination and, 153–154
    segment reassortment in, 155
    food-borne. *See* Food-borne viruses
    genetic variation of, molecular mechanisms of, 150–155
    growth of, compounds in foods and, 196
    hepatitis, enterically transmitted, 80–81
    inactivation of, 12–13, 19
        and persistence of, 195–198
        cooking for, 199–201
        effects of environmental factors on, 196–197
        effects of temperature on, 195–196
        factors influencing, 191, 195
        high-hydrostatic-pressure processing for, 198–199
        in shellfish, physical factors and, 200–201
    mammalian, taxonomic groupings of, 124, 125
    media for culture of, 3–4
    microevolution of, 150
    origin of word, 1
    persistence in environment, 163–164, 226
    persistence in foods, consequences of, 201–202
    persistence in shellfish, 197–198
    prevention of transmission of, 15–19
    propagation of, 1, 3–5
    proteins of, genes encoding, 149–150
    replication of, 5
    subpopulations of, selection in environment, 163–164
    transmission of, cross-species, 158
        secondary, risk assessment models and, 231
    transmission via food, 156–157
        factors influencing, 189
    treatment of foods to remove or inactivate, 198–201
    uptake by shellfish, 193–195
    vessels for culture of, 3

W
Wastewater, antiviral treatment of, 17
    fecal-oral transmission via, 17, 18
Water, contamination of, 176
    irrigation, transfer of viruses from, 230–231
    protozoa in, model for estimation of, 222
West Nile virus (WNV), 132

World Health Organization (WHO), food-
    borne illnesses and, 209
  role in risk assessment, 214–215
  World Health Assembly 1953 statement,
    213
World Trade Organization (WTO),
    food-related illness and, 211–212

Y
Yellow fever virus, 132

Z
Zoonoses, 122
Zoonotic virus, transmission via food,
    157–158